SIMULATED ANNEALING FOR VLSI DESIGN

THE KLUWER INTERNATIONAL SERIES IN ENGINEERING AND COMPUTER SCIENCE

VLSI, COMPUTER ARCHITECTURE AND DIGITAL SIGNAL PROCESSING

Consulting Editor

Jonathan Allen

Other books in the series:

Logic Minimization Algorithms for VLSI Synthesis. R.K. Brayton, G.D. Hachtel, C.T. McMullen, and A.L. Sangiovanni-Vincentelli. ISBN 0-89838-164-9.
Adaptive Filters: Structures, Algorithms, and Applications. M.L. Honig and D.G. Messerschmitt. ISBN 0-89838-163-0.
Computer-Aided Design and VLSI Device Development. K.M. Cham, S.-Y. Oh, D. Chin and J.L. Moll. ISBN 0-89838-204-1.
Introduction to VLSI Silicon Devices: Physics, Technology and Characterization. B. El-Kareh and R.J. Bombard. ISBN 0-89838-210-6.
Latchup in CMOS Technology: The Problem and Its Cure. R.R. Troutman. ISBN 0-89838-215-7.
Digital CMOS Circuit Design. M. Annaratone. ISBN 0-89838-224-6.
The Bounding Approach to VLSI Circuit Simulation. C.A. Zukowski. ISBN 0-89838-176-2.
Multi-Level Simulation for VLSI Design. D.D. Hill and D.R. Coelho. ISBN 0-89838-184-3.
Relaxation Techniques for the Simulation of VLSI Circuits. J. White and A. Sangiovanni-Vincentelli. ISBN 0-89838-186-X.
VLSI CAD Tools and Applications. W. Fichtner and M. Morf, editors. ISBN 0-89838-193-2.
A VLSI Architecture for Concurrent Data Structures. W.J. Dally. ISBN 0-89838-235-1.
Yield Simulation for Integrated Circuits. D.M.H. Walker. ISBN 0-89838-244-0.
VLSI Specification, Verification and Synthesis. G. Birtwistle and P.A. Subrahmanyam. ISBN 0-89838-246-7.
Fundamentals of Computer-Aided Circuit Simulation. W.J. McCalla. ISBN 0-89838-248-3.
Serial Data Computation. S.G. Smith and P.B. Denyer. ISBN 0-89838-253-X.
Phonological Parsing in Speech Recognition. K.W. Church. ISBN 0-89838-250-5.

SIMULATED ANNEALING FOR VLSI DESIGN

by

D.F. Wong
University of Texas at Austin

H.W. Leong
National University of Singapore

C.L. Liu
University of Illinois at Urbana-Champaign

KLUWER ACADEMIC PUBLISHERS
Boston/Lancaster/Dordrecht

Distributors for North America:
Kluwer Academic Publishers
101 Philip Drive
Assinippi Park
Norwell, Massachusetts 02061, USA

Distributors for the UK and Ireland:
Kluwer Academic Publishers
MTP Press Limited
Falcon House, Queen Square
Lancaster LA1 1RN, UNITED KINGDOM

Distributors for all other countries:
Kluwer Academic Publishers Group
Distribution Centre
Post Office Box 322
3300 AH Dordrecht, THE NETHERLANDS

Library of Congress Cataloging-in-Publication Data

Wong, D.F.
Simulated annealing for VLSI design / by D.F. Wong, H.W. Leong, C.L. Liu.
p. cm. — (Kluwer international series in engineering and computer science ; SECS 42)
Bibliography: p.
Includes index.
ISBN-13: 978-1-4612-8947-0 e-ISBN-13: 978-1-4613-1677-0
DOI: 10.1007/978-1-4613-1677-0
1. Integrated circuits—Very large scale integration—Design and construction—Data processing. 2. Annealing of metals—Mathematical models. I. Leong, Hon Wai, 1955- . II. Liu, C.L. (Chung Laung), 1934- . III. Title. IV. Series.
TK7874.W65 1988 87-29306
621.381'73—dc19 CIP

CONTENTS

PREFACE

This monograph represents a summary of our work in the last two years in applying the method of simulated annealing to the solution of problems that arise in the physical design of VLSI circuits. Our study is experimental in nature, in that we are concerned with issues such as solution representations, neighborhood structures, cost functions, approximation schemes, and so on, in order to obtain good design results in a reasonable amount of computation time. We hope that our experiences with the techniques we employed, some of which indeed bear certain similarities for different problems, could be useful as hints and guides for other researchers in applying the method to the solution of other problems.

Work reported in this monograph was partially supported by the National Science Foundation under grant MIP 87–03273, by the Semiconductor Research Corporation under contract 87–DP–109, by a grant from the General Electric Company, and by a grant from the Sandia Laboratories.

We gratefully acknowledge Professor Xianjin Yao's assistance in the preparation of the monograph.

SIMULATED ANNEALING FOR VLSI DESIGN

CHAPTER 1

INTRODUCTION

1.1. Combinatorial Optimization

For many important practical or theoretical problems, the objective is to choose a "best" solution out of a large number of possible solutions. Such problems are typically known as *combinatorial optimization* problems [Lawl76, PaSt82]. In many combinatorial optimization problems, a solution is an arrangement of a set of discrete objects according to a given set of constraints. A solution is also called a *configuration*. The set of all solutions is referred to as the *solution space*. A cost function f is defined on all solutions. That is, for a solution x, $f(x)$ is the cost of the solution. Our goal is to develop efficient algorithms for determining a configuration that *minimizes* the value of the cost function. An example of combinatorial optimization problems is the well–known Traveling Salesman Problem [PaSt82] which is the problem of determining a *traveling salesman route* (a route which passes through each of a given set of cities once and returns finally to the starting city) with minimum cost. In this case, the cities are the discrete objects to be rearranged. That each city must be visited once and only once is the constraint which must be satisfied.

Significant progress has been made recently in the study of combinatorial optimization. Important problems such as the Maximum Matching Problem and the Maximum Flow Problem [Lawl76, PaSt82] have been shown to be solvable in polynomial time. That is, polynomial time algorithms (*i.e.* algorithms which require computing efforts proportional to a polynomial of the problem size) that determine optimal solutions for these problems have been discovered. Furthermore, recent research in design of data structures and analysis of algorithms [AhHU74, ReND77, Tarj83] lead to more efficient implementations of many combinatorial optimization algorithms. However, for many other important combinatorial optimization problems (*e.g.* the Traveling Salesman Problem) encountered in practice, efficient algorithms for determining optimal solutions are still unknown. In fact, all known algorithms for these problems require a computing effort that increases exponentially with the problem size. Study of the notion of NP–completeness [GaJo79] suggests that it is unlikely that polynomial time algorithms will be found for the solution of these problems. Since many of the NP–complete problems are of practical importance, efficient approximation algorithms, namely, algorithms that do not produce optimal but rather close–to–optimal solutions, have been developed.

A general strategy for designing approximation algorithms is the method of *iterative improvement* (or called *local search* in [PaSt82]). This method is based on the simple and natural technique of trial and error. Iterative improvement has been shown to be effective for solving many difficult combinatorial optimization problems [KeLi70, LiKe73]. Application of the method of iterative improvement requires the definition of a solution space, a cost function, and a set of *moves* that can be used to modify a solution. A solution y is a *neighbor* (*neighboring solution*) of a solution x if y can be obtained from x via one of the moves. In the method of iterative improvement, one starts with an initial solution and examines its neighbors until a neighboring solution y with a lower cost is discovered. In that case, the solution y becomes the new solution and the process is continued to examine the neighbors of the new solution. The algorithm terminates when it arrives at a solution which has no neighboring solution with a lower cost. The method of simulated annealing to be studied in this monograph is an extension of the method of iterative improvement based on an analogy between a combinatorial optimization problem and the problem of determining the lowest–energy ground state of a physical system.

1.2. The Method of Simulated Annealing

Statistical Mechanics is the study of the behavior of complex systems consisting of a large number of interacting atoms in thermal equilibrium at a finite temperature. In particular, the atomic states at any temperature T satisfy the Boltzmann's distribution, namely, the probability that a system is in a given state r is given by $e^{E(r)/k_b T}$ where $E(r)$ is the energy associated with state r, and k_b is the Boltzmann's constant. Hence, under equilibrium, the most probable states at any given temperature are those with lowest energy.

The analogy between a combinatorial optimization problem and the problem of determining the lowest–energy ground state of a physical system with many interacting atoms was first observed by Kirpatrick, Gelatt and Vecchi [KiGV83] and Cerny [Cern85]. This analogy is summarized in Fig.1.1. The ***states*** of a system correspond exactly to the ***configurations*** of the combinatorial optimization problem. The ***ground states*** of the system correspond to *optimal configurations*, namely, configurations that minimize the cost function. Finally, the problem of determining an optimal configuration in combinatorial optimization corresponds to that of determining a low–temperature ground state of fluids. Based on these observations, Kirpatrick, Gelatt, and Vecchi proposed to apply Monte–Carlo methods, which are traditionally used in statistical mechanics, to the solution of combinatorial optimization problems.

It is well–known that to bring a fluid to a highly ordered, low–energy state (for example, in growing a single large crystal), a process called *careful annealing* can be employed. We first melt

Physical Systems	**Optimization Problems**
State	*Configuration*
Energy	*Cost Function*
Ground State	*Optimal Solution*
Rapid Quenching	*Iterative Improvement*
Careful Annealing	*Simulated Annealing*

Fig. 1.1 Analogy between physical systems and optimization problems.

the system by heating it up to some high temperature, and then cool it slowly, spending a long time at temperatures in the vicinity of the freezing point. At each temperature during the annealing process, slow cooling enables the system to achieve equilibrium. If the temperature is lowered too quickly, the system does not have sufficient time to achieve equilibrium, and the resulting configuration might have many defects in the form of high–energy, metastable, locally optimal structures. Intuitively, fast cooling in physical systems corresponds to an iterative improvement scheme in combinatorial optimization. Kirpatrick *et al* suggested that better results to combinatorial optimization problems can be obtained by simulating the annealing process of physical systems.

In 1953, Metropolis *et al* [MeRR53] introduced a computational procedure for efficient simulation of the equilibrium states of a many–body system at a given finite temperature. Metropolis' procedure is as follows: In each step, a small perturbation of the configuration is chosen at random and the resulting change in the energy of the system, Δ, is computed. The new configuration is accepted with probability 1 if $\Delta \leq 0$, and with probability $e^{-\Delta/k_b T}$ if $\Delta > 0$. This procedure is shown in Fig.1.2. The procedure can be easily adapted for the solution of optimization problems as follows: Replace states in the physical system by configurations in the optimization problem; replace the energy function of a given state

Metropolis' Procedure

begin
    Choose some random initial configuration S ;
    repeat
        S' := Some random neighboring configuration of S ;
        $\Delta := E(S') - E(S)$;
        $Prob := \min(1, e^{-\Delta/k_b T})$;
        if $random(0,1) \leq Prob$ **then** $S := S'$;
    until false;
end;

Fig. 1.2 Metropolis' procedure.

by the cost function of the corresponding configuration. Then, this procedure can be used to simulate the behavior of the combinatorial problem with the given cost function.

Kirpatrick *et al* generalized this basic approach by introducing a multi–temperature approach in which the "temperature" is lowered slowly in stages. At each temperature, the system is simulated by Metropolis' procedure until the system reaches equilibrium. The generic simulated annealing optimization procedure is given in Fig.1.3. Note that in the simulated annealing procedure for combinatorial optimization, the Boltzmann's constant is combined with the temperature and we shall use the term "temperature" to refer to their product. Temperature can also be viewed as nothing but a control parameter for the optimization procedure.

Intuitively, the simulated annealing optimization procedure can be viewed as an enhanced version of iterative improvement.

Generic Simulated Annealing Algorithm

```
begin
   S := Initial solution S_0 ;
   T := Initial temperature T_0 ;
   while (stopping criterion is not satisfied) do
      begin
         while (not yet in equilibrium) do
            begin
               S' := Some random neighboring solution of S ;
               Δ := C(S') - C(S) ;
               Prob := min ( 1, e^(-Δ/T) ) ;
               if random(0,1) ≤ Prob then S := S' ;
            end ;
         Update T ;
      end ;
   Output best solution ;
end ;
```

Fig. 1.3 Generic simulated annealing algorithm.

The simulated annealing procedure randomizes the iterative improvement procedure and also allows occasional "uphill moves" (moves that worsen the current solution) in an attempt to reduce the probability of being stuck at a locally optimal solution. These uphill moves are controlled probabilistically by the temperature T, and become less and less likely toward the end of the process, as the value of T decreases.

Theoretical studies of the simulated annealing optimization technique have been reported in [AaLa85, GeMi85, Haje85, LuMe86, RoSa85, MiRS86] where Markov chains [Fell50] are used as the mathematical model. Theoretical analysis shows that this class of algorithms converges with probability 1 to a globally optimal solution, provided certain conditions on the number of iterations for each temperature T, and a certain rule for updating the value of T is followed. These convergence results are, unfortunately, asymptotic and provide little information on how to choose the various parameters in the implementation of a particular algorithm. Consequently, most of the current applications of the method of simulated annealing employ the simple and yet effective approach of [KiGV83] or its variations to choose these parameters. An important current research topic on simulated annealing is that of determining good estimates of these parameters so that there will be more theoretical guidelines for a more effective implementation of simulated annealing algorithms [AaLa85, HuRS86, LuMe86, OtGi84, Whit84]. Nevertheless, the convergence results give confidence on the well-posedness of the method of simulated annealing, and help to account for the success of the simulated annealing approach in practice. We refer the readers to [LaAa87] for an introduction to many of the theoretical results of simulated annealing.

The method of simulated annealing as an optimization technique is very appealing because it produces high quality solutions and is in general easy to implement. It must be noted that simulated annealing is a general design methodology rather than a completely specified algorithm. Thus, application of the method of simulated annealing to a particular optimization problem requires careful design of the basic ingredients: (1) formulating the problem so as to obtain a concise description of the configurations, (2) defining the neighboring solutions of each solution, (3) choosing a suitable cost function, and (4) defining an *annealing schedule* (*i.e.*

specifying the initial temperature, the rule for changing the temperatures, the duration of search at each temperature, and the termination condition of the algorithm).

1.3. Remarks

Recent research has demonstrated the effectiveness of simulated annealing in the solution of many optimization problems in the area of design automation (for VLSI circuits). In particular, the method of simulated annealing has been successfully applied to problems including *placement* [Grov86, JeGe83, KiGV83, SeSa85], *floorplan design* [OtGi84, WoLi86, WoLi87a, Wong87], *routing* [VeKi83, LeWL85, LeLi87, Leon86a], *PLA folding* [MoGe85, WoLL86, WoLL87], *Gate Matrix layout* [Leon86b], *logic array optimization* [DeNe86a, DeNe86b, WoLi87], *two-dimensional compaction* [Zees85, Most86, Osma87, HsLL87], *via minimization* [Moli85], *logic minimization* [FlGM85, LaDe86, Gons86, YaLi87], *testing* [DiPi86, LiAB86], *transistor sizing* [PiDe86], and *digital filter design* [CaDV85, CaDV86]. Also, many researchers have worked on parallel implementations of simulated annealing. Most of these algorithms are for placement [CaRS86, BaJo86, RoBS86, DaKN87, FiWa87, KrRu87, RaPa87] but there are also some for other problems [ChRa86, DeNe86].

In the remaining chapters, we shall present simulated annealing based algorithms for placement, floorplan design, channel routing, PLA folding, Gate Matrix layout, and general logic array optimization.

CHAPTER 2

PLACEMENT

2.1. Introduction

The general placement problem is the problem of placing a set of circuit modules on a chip such that a certain objective function is minimized. The ultimate goal is to minimize the total chip area occupied by the circuit modules and the interconnection between the modules. To make the placement problem computationally feasible, various objective functions based on parameters such as the area of the bounding rectangle that contains all the modules, the total interconnection wire length, or some other kind of routing area estimation, are commonly used. Placement problems can be classified according to the different types of design methodology. The three types of placement problems are *gate–array placement* [HaWA73, QuBr79, KiGV83, FrKa86], *standard–cell placement* [Breu77, DuKe85, SeSa86, SuKe86], and *macro/custom–cell placement* [PrVa79, Laut80, JeGe83, ShDu85].

A gate array is a two–dimensional matrix of identical cells, each containing a fixed number of uncommitted transistors separated by wiring channels. (See Fig.2.1.) Each cell can be used to implement a logical gate by appropriately specifying the interconnections among the transistors within the cell. In the gate–array design methodology, a circuit is formed by first selecting a set of cells in the gate array for the implementation of the required logical gates, and then interconnecting these logical gates using the wiring channels between the cells. The gate–array placement problem is that of selecting a set of cells to implement the logical gates. We can view this problem as a problem of placing a set of uniformly sized circuit modules onto a matrix of cells. This placement problem is very similar to the classical PCB (printed circuit board) placement problem.

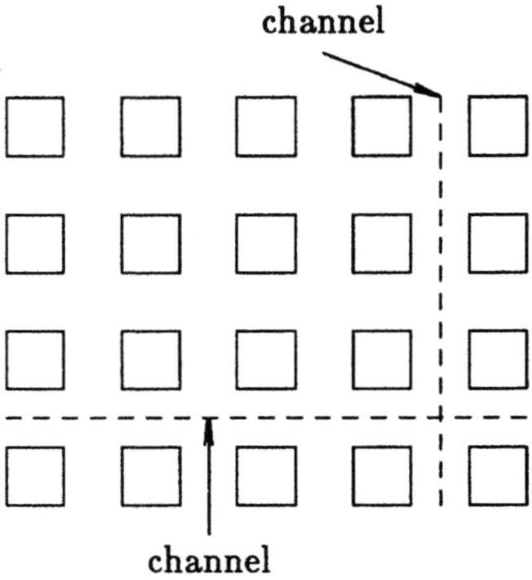

Fig. 2.1 Gate–array placement.

In the standard–cell design methodology, the designer constructs a circuit by selecting modules from a library of modules that implement pre–defined logic functions. These modules, usually referred to as standard cells or polycells, were pre–designed and their dimensions, performances, and electrical characteristics have already been determined. In general, standard cells are of the same height but varying width. Thus, they are typically fit together in rows. Each standard cell has terminals on its top and/or bottom. The terminals are connected by wires that run in channels between and around the rows of standard cells. Feedthrough cells may be used to make connections that span more than one row. The standard–cell placement problem is that of arranging a set of standard cells in horizontal rows. (See Fig.2.2.)

Semi–custom design methodologies such as gate–array and standard–cell have the advantage of reducing design time but at the expense of chip area. As the complexity of VLSI circuits increases, effective CAD tools that can support full custom design become increasingly important. The macro/custom–cell placement problem is that of placing circuit modules of different sizes on a chip. (See Fig.2.3.) This problem is more difficult than gate–array or standard–cell placement because in this case it is more difficult to estimate the routing area due to irregularities in module size.

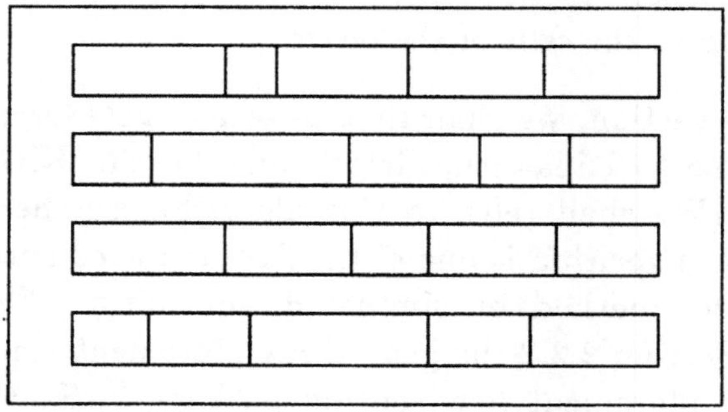

Fig. 2.2 Standard–cell placement.

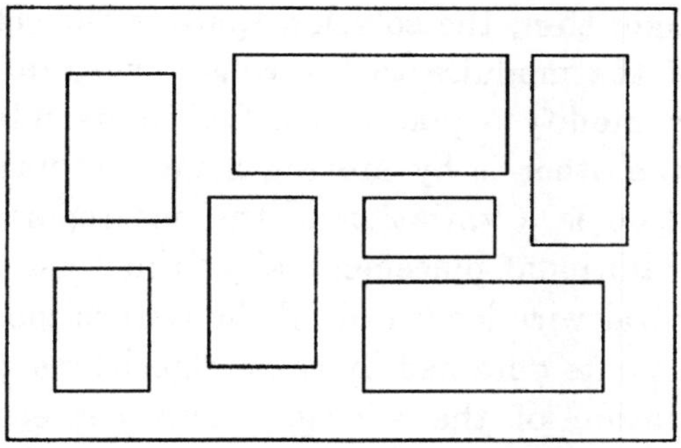

Fig. 2.3 Macro/custom–cell placement.

In this chapter, we present simulated–annealing based algorithms as well as other similar stochastic algorithms for these three types of placement problems. In Section 2.2, we describe algorithms for gate–array placement. Section 2.3 presents algorithms for standard–cell placement. Section 2.4 describes algorithms for custom–cell placement. Finally, in Section 2.5, we present some other similar stochastic approaches for the placement problems.

2.2. Gate–Array Placement

Let **M** be a set of n given modules. Let $\mathbf{N} = \{N_1, N_2, ..., N_m\}$ be a set of m signal nets where each net N_i, $N_i \subseteq \mathbf{M}$, is a set of modules to be interconnected. We shall represent a gate array by a two–dimensional board in which vertical lines correspond to the vertical channels and horizontal lines correspond to the horizontal channels. A placement is thus an assignment of the modules in **M** to the cells of the board.

In this section, we shall first present a gate-array placement algorithm due to Kipatrick, Gelatt, and Vecchi [KiGV83] in Section 2.2.1. (We shall refer to this algorithm as the K-G-V algorithm.) This algorithm is one of the first reported successful application of the method of simulated annealing. Then we shall describe in Section 2.2.2 the gate-array placement algorithm in the TimberWolf placement and routing package [SeSa85]. This algorithm is a refinement of the K-G-V algorithm.

2.2.1. The K-G-V Algorithm

In this algorithm, the solution space is the set of all possible assignments of the modules to the cells. Two types of moves are used to locally modify a placement. One is by interchanging two modules and the other is by moving a module into a vacant cell. The cost function is a variation of the one reported in [Souk81]. Most of the traditional placement algorithms use a cost function based on the total wire length of all the connections. However, the resulting placements obtained by these algorithms might have very uneven distribution of the wirings. This consequently leads to crowded unroutable channels in the gate array. We shall use a cost function that is based on the concept of *net-crossing histogram.* This cost function tends to distribute the wirings more evenly among the channels and consequently improves the routability of the placement.

For a placement of the modules, let S_i be the set of cells in which the modules of net N_i are placed. Let R_i be the smallest rectangle enclosing the centers of the cells in S_i. The rectangle R_i is called the *bounding rectangle* of net N_i. Let L_1, L_2, ..., L_k denote all the channels (vertical or horizontal) in the gate array. For each channel L_i, let w_i be the total number of nets whose bounding rectangles intersect with channel L_i. (In [KiGV83], the set of w_i's for all vertical channels L_i is called the *vertical net-crossing histogram*. Also, the set of w_i's for all horizontal channels L_i is called the *horizontal net-crossing histogram.*) Let $\mathbf{W} = \sum_{i=1}^{k} w_i$. Note that **W** can also be obtained by summing the half perimeter of the bounding rectangle R_i of each net N_i. Hence **W** is an estimation of the total wire length. In order to measure routability of a placement, we introduce a threshold value t_i for each channel L_i. The threshold values are the wiring capacities of the channels. For each channel L_i, let $\delta_i = w_i - t_i$ if $w_i > t_i$, and $\delta_i = 0$ otherwise. Let $\Delta = \sum_{i=1}^{k} \delta_i^2$. The cost function is

$$\Psi = \mathbf{W} + \alpha \cdot \Delta$$

where α is a constant. The term Δ has the effect of distributing the wirings more evenly among the channels.

Finally, the temperature schedule used is of the form $T_i = r \cdot T_{i-1}$, $i = 1, 2, 3, \ldots$, where r is a constant close to 1 (*e.g.* $r = 0.9$). Experimental result for an example of a placement problem with 98 modules on a 10×10 gate array is reported in [KiGV83]. They started with an initial placement in which each group of functionally related modules are placed close together by a human designer. In that placement, wiring in the middle portion of the chip is much more congested than at other places. At high temperature, the initial placement was unscrambled into a random–looking placement. In the final placement, functionally related modules were grouped together again but there was no more wiring congestion. The maximum histogram values were about 30% less than that in the initial placement. Also the the total wire length was decreased by about 10%. About 250,000 interchanges were attempted. It took 12 minutes of computation on an IBM 3033.

2.2.2. TimberWolf

The TimberWolf placement and routing package [SeSa85] has a program for gate–array placement that is similar to that described above. In this program, the user can choose to use either one of two cost functions. The first cost function is the same as that in [KiGV83]. The second cost function is a refinement of the first one. It examines the local wiring congestion more closely. We shall refer the point of intersection between a vertical channel and a horizontal channel as a *cut point*. Each channel (vertical or horizontal) is partitioned into a set of *channel segments* by the cut points on the channel. (See Fig.2.4.) Let $L_1^{'}$, $L_2^{'}$, ..., $L_l^{'}$ denote all

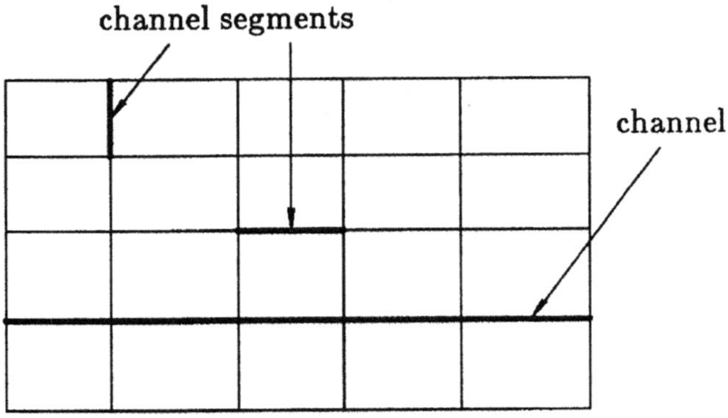

Fig. 2.4 Channel segments.

channel segments in the gate array. We assign a threshold value t_i to each channel segment L_i'. For each net N_j, let Q_j be the set of all channel segments that are intersected or enclosed by the bounding rectangle R_j. Let p_j be the half perimeter of R_j. The congestion per channel segment in Q_j introduced by net N_j is approximately p_j/l_j where l_j is the number of channel segments in Q_j. The term p_j/l_j can also be interpreted as the expected occupancy for net N_j in each of the channel segments in Q_j. For each channel segment L_i' and net N_j, let $w_{ij} = p_j/l_j$ if L_i' is in Q_j, and let $w_{ij} =$ 0 otherwise. Let $w_i = \sum_{j=1}^{m} w_{ij}$. The term w_i is an estimate of the expected occupancy in channel segment L_i' by all nets. For each channel segment L_i', let $\delta_i = w_i - t_i$ if $w_i > t_i$, and $\delta_i = 0$ otherwise. The cost function is

$$\Psi = \sum_{i=1}^{l} w_i + \alpha \sum_{i=1}^{l} \delta_i^2$$

where α is a constant. In other words, this cost function is the sum of the expected occupancies of the segments plus a penalty assessed to each channel segment which has occupancy exceeding the corresponding threshold.

Note that so far in our construction of the bounding rectangles of the nets, we assume that the pin terminals associated with a net are in the center of each of the modules involved. In fact, TimberWolf allows the user to specify on which side a pin terminal lies.

Let S_i be the set of cells in which the modules of net N_i are placed. For each cell A in S_i, we *mark* the mid-point of an edge of A if the module placed in A has a pin terminal belonging to net N_i and is on the same side of that edge. The bounding rectangle R_i of net N_i is then defined as the smallest rectangle enclosing all the marked points for net N_i. With the new definition of the bounding rectangle, we have a cost function that can measure more accurately the quality of a placement.

Experimental results were given for three placement problems (ILLIAC IV computer boards) previously considered by Stevens [Stev72]. The largest example is one in which 150 modules are to be placed on a 11 $\times$ 15 board. TimberWolf reduced the total estimated wire length by 21% over Steven's result, and by 17% over the result reported by Goto and Kuh [GoKu78]. Total computation time was 15 CPU minutes on a VAX 11/780 computer.

2.3. Standard-Cell Placement

In this section, we shall present two different standard-cell placement algorithms. The first is the standard-cell placement algorithm in the TimberWolf placement and routing package [SeSa85, SeSa86], and the second is due to Grover [Grov86]. These two algorithms will be described in Sections 2.3.1 and 2.3.2, respectively. A major difference between these two algorithms is the definition of the solution space. TimberWolf allows placements with overlapping modules as intermediate solutions. Consequently, the cost function can be updated efficiently after every move. On the other hand, the algorithm in [Grov86] does not allow placements with overlapping modules as intermediate solutions. In this case, it is in general more expensive to update the cost function after a move. One possible advantage of this approach is the reduction of the size of the solution space which in general reduces search efforts.

2.3.1. TimberWolf

The TimberWolf standard-cell placement algorithm consists of two stages. In stage 1, modules are moved between different rows as well as within the same row. Module overlaps are allowed in this stage. When the temperature is reduced below a certain value, stage 2 begins. First, all module overlaps are removed by the following procedure: For each row, the modules are ordered in increasing order of the x−coordinates of their centers. They are then placed side by side starting from the left edge of the row. After that, the annealing process continues but only adjacent modules are to be interchanged. Modules will no longer be moved between different rows. We now describe the details of the algorithm.

Solution Space

A ***placement solution*** is an arrangement of the given modules into rows possibly with overlaps. (See Fig.2.5.) In stage 1, the solution space consists of all placement solutions, whereas in stage 2, only placement solutions with no module overlaps are allowed.

Neighboring Solutions

Three types of moves are used to locally modify a placement solution. These three types of moves are:

M1. Displace a module to a new location.
M2. Interchange two modules.
M3. Change the orientation of a module.

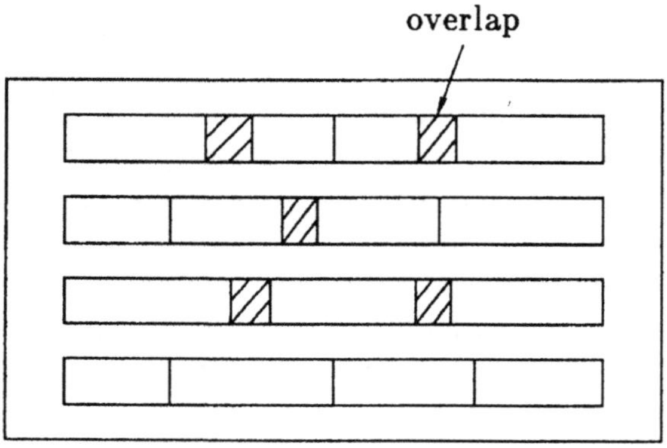

Fig. 2.5 Module overlaps.

A move of type M3corresponds to a reflection with respect to the horizontal or the vertical line through the center of the module. This move is needed because the cost function depends on the locations of the pin terminals. Note that a move of type M1 may move a module to a different row. Also note that a move of type M2 may introduce overlap of modules because the width of the modules in general are different. Given a placement solution, a move is chosen to modify this solution by making a probabilistic choice among M1, M2, and M3. TimberWolf always first try to select a move between type M1 and M2 with probability 4 times higher in favor of M1. If a move of type M1 is chosen (for a certain module) and the new placement solution is rejected, then a move of type M3 (for the same module) will be chosen with probability 1/10.

There are restrictions on how far a module can be displaced by a move of type M1 and what pairs of modules can be interchanged by a move of type M2. A *range limiter* is a rectangular window R. For a move of type M1, the window R is centered at the center of the module to be displaced. A random choice within R will be made as the new location for the module. For a move of type M2, two modules can be interchanged only if the window R can be positioned such that it contains the centers of both modules. At the beginning of the annealing process, the window size is set to be large enough to contain the whole chip, and it shrinks slowly as the temperature decreases. In fact, the height and width of the window are proportional to the logarithm of the temperature T.

Stage 2 begins when the temperature is reduced to a point at which the window size becomes so small that modules are no longer permitted to switch rows. At this time, all module overlaps are removed by the procedure described earlier. Then the annealing process continues but only moves of type M2 (only those that swap adjacent modules) or M3 are allowed. Again, a move of type M3 is attempted only when a move of type M2 is attempted and rejected.

Cost Function

Let N_1, N_2, ..., N_m denote all the signal nets. For each net N_i, let R_i be the smallest rectangle enclosing all the pin terminals that belong to the net in a placement solution. Let w_i be the width and h_i be the height of R_i. The cost function is:

$$\Psi = \mathbf{C}_1 + \mathbf{C}_2 + \mathbf{C}_3$$

where $\mathbf{C}_1$, $\mathbf{C}_2$, and $\mathbf{C}_3$ are three different functions that are based on the total interconnection cost, the amount of module overlaps, and the sum of deviations of row lengths from their desired values, respectively.

The first function $\mathbf{C}_1$ is an estimation of the total interconnection cost. It is defined as follows:

$$\mathbf{C}_1 = \sum_{i=1}^{m} (\alpha_i w_i + \beta_i h_i)$$

where α_i and β_i are user specified *horizontal* and *vertical weights* for net N_i. (For the case $\alpha_i = \beta_i = 1$ for all net N_i, $\mathbf{C}_1$ is the typical total wire length estimation obtained by summing up the half perimeters of the bounding rectangles of all the nets.) For critical nets, it is customary to increase both the vertical and horizontal weights of the nets in order to ensure that these nets will be kept as short as possible. Also, in the case of double metal technology, many connections that span more than one row can be made by over–the–cell routing and consequently fewer of feedthrough cells are needed. In this case, vertical wirings are in some sense cheaper than horizontal wirings. Thus, we should set the vertical weights smaller than the horizontal weights in order to achieve better placement results.

The second function $\mathbf{C}_2$ is a penalty function for module overlaps. Since module overlaps are not acceptable in the final placement solution, a penalty function that measures the amount of overlap of the modules is used to discourage module overlaps. For every pair of modules i and j, let $O(i,j)$ be the amount of overlap between the two modules in the x–dimension. The function $\mathbf{C}_2$ is defined as follows:

$$\mathbf{C}_2 = \sum_{i \neq j} (O(i,j) + \alpha)^2$$

where α is an offset parameter added to ensure that $\mathbf{C}_2$ converges to 0 when temperature approaches 0.

The third function $\mathbf{C}_3$ is a penalty function that controls the row lengths. For each row r, there is a desired row length $d(r)$. Let $l(r)$ be the sum of the widths of the modules in row r. The function $\mathbf{C_3}$ is defined as follows:

$$\mathbf{C_3} = \sum_r \beta \cdot |l(r) - d(r)|$$

where β is a user specified constant.

Note that in stage 2, modules do not change rows and there are no module overlaps. Thus $\mathbf{C_2}$ is 0 and $\mathbf{C_3}$ remains constant throughout this stage. The cost function is effectively just $\mathbf{C_1}$.

Annealing Schedule

The temperature schedule is of the form $T_k = r(k) \cdot T_{k-1}$ for $k = 1, 2, 3, \ldots$. The values of $r(k)$ increases from 0.8 to a maximum value 0.94 and then decreases to 0.1. At each temperature, the total number of moves attempted is $K \cdot n$ where n is the total number of modules and K is a user specified constant.

Experimental Results

The TimberWolf placement program was compared with the CIPAR standard cell placement package developed by American Microsystems, Inc. For the larger circuits tested (800 to 2700 modules), TimberWolf achieved total estimated wire length reduction ranging from 45% to 66%. As for the running time, for the largest circuit tested (2700 modules), 75 million moves were performed. The computation time was 6.5 CPU hours on an IBM 3081 computer (equivalent to approximately 84 CPU hours on a Vax 11/780 computer).

It should be noted that an efficient general annealing schedule [HuRS86] was proposed by Huang, Romeo, and Sangiovanni–Vincentelli recently. They have installed the new schedule in the TimberWolf standard cell placement package. Experimental results show that savings in CPU time up to 57% was possible.

2.3.2. Another Approach

In [Grov86], Grover proposed a different annealing algorithm for the standard–cell placement problem. The two main features of his approach are: (1) solutions in the solution space do not allow module overlaps; and (2) the cost calculation is *approximate* instead of being *exact*. A *placement solution* is an arrangement of the modules into rows such that the modules in each row are placed adjacent to each other. There is neither overlap nor empty space between adjacent modules. Two types of moves are used to locally modify a placement solution. The first is by removing a module from a row and inserting it either at the beginning, or at the end, or between two adjacent modules in some row. After a module is removed from a row, all the modules to its right are shifted correspondingly so that no empty space is left. When a module is inserted into a row, all the modules to the right of its new location will be shifted right to make room for the module. (See Fig.2.6.) The second type of move is to swap two modules. Again, if necessary, modules to the right of the new location of each of the two modules will be shifted either to the right or to the left because the two modules are, in general, of different widths. (See Fig.2.7.)

The cost function used in [Grov86] is the standard total wire length estimation which is the sum of the half perimeters of the bounding rectangles of all the nets. This corresponds to the function $\mathbf{C}_1$ in TimberWolf with $\alpha_i = \beta_i = 1$ for all net N_i. (Of course, if desired, $\mathbf{C}_1$ can also be used as the cost function in Grover's algorithm.) Since each step in the annealing process might involve movement of many modules, it is in general more expensive to update the cost function than in the case of TimberWolf. In order to reduce the computation time, cost calculations are carried out *approximately* by only updating the positions of the modules every U moves. Let $\Delta\Psi$ be the change in cost and T be

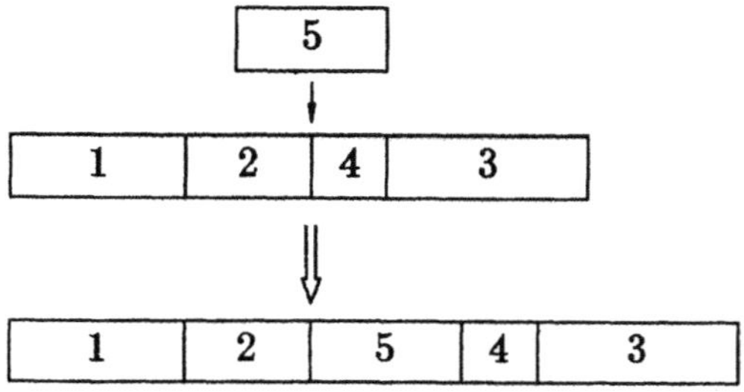

Fig. 2.6 Insert a module into a row.

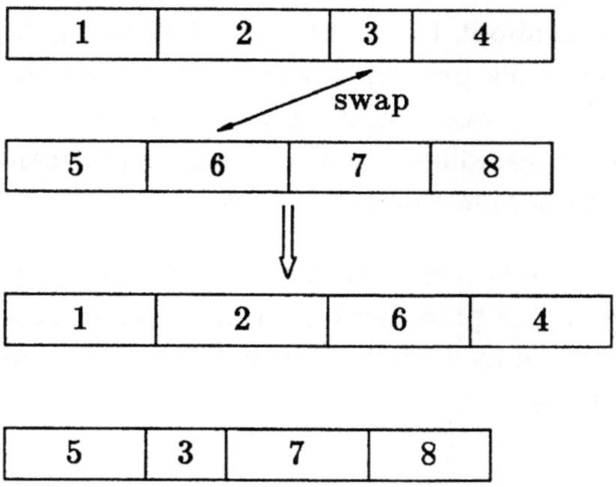

Fig. 2.7 Interchange two modules.

the temperature. Accordingly, the probability of accepting an uphill move is $\exp(-\Delta\Psi/T)$. This probability will not be significantly changed if $\Delta\Psi$ is replaced by $\Delta\Psi \pm \delta$ for $\delta \ll T$. Thus, as long as the error in the cost calculation is small compared with the temperature, we could expect that the algorithm is not significantly disturbed. In Grover's algorithm, the positions of the modules are updated every U moves. The error in the cost calculation is due to the inaccurate information of the positions of the modules. It was shown in [Grov86] that as long as

$$U < R\frac{\alpha^2 T^2}{2mw^2}$$

the expected error of the cost calculation is less than $\alpha \cdot T$, where T is the temperature, m is the total number of nets, R is the number of rows, and w is the average width of the modules. It was also suggested that $\alpha = 1$ is a reasonable choice. For a typical circuit: $w = 75\mu$, $R = 20$, $m = 1500$. A typical initial temperature is $T = 10^6\mu$. This implies that for $\alpha = 1$, we only have to update the positions of the modules once every 10^6 moves at the beginning. This is reasonable because almost all moves are accepted at very high temperature.

The algorithm was compared with a standard cell placement program in LTX2, an AT&T Bell Labs VLSI layout system, on 8 problems of size ranging from 1000 to 4000 standard cells.

Improvements of about 20–40% in routing area, leading to a 10–20% reduction in chip area, were observed. The running time of the program was about 1.5 to 2 CPU hours on an Amdahl 5870 for a 3000 standard cells problem, whereas the running time of a version that performs exact cost calculation after every move was more than 5 times longer, without any significant difference in quality of the final placements obtained.

Finally, Grover reported in [Grov87] that if a good initial solution is used, this placement program can start at a lower temperature and arrive at a final solution with comparable quality in a much shorter time.

2.4. Macro/Custom–Cell Placement

We shall describe in Section 2.4.1 an algorithm developed by Jespen and Gelatt [JeGe83], and in Section 2.4.2 the macro/custom–cell placement algorithm of TimberWolf [SeSa85].

2.4.1. Jespen and Gelatt's Algorithm

The discussion here is based on [JeGe83]. A *placement solution* is specified by the locations of all the modules on the chip as well as their rotations and reflections from standard orientations. All modules are rectangular and they are in general of varying sizes and shapes. As in standard–cell placement, module overlaps are allowed in a placement solution. This greatly simplifies the definitions of the moves. Three types of moves are used to locally modify a placement. They are:

M1. A horizontal or vertical translation of a module.
M2. A 90° rotation in either direction.
M3. A horizontal or vertical reflection.

Note that the first two types of moves in general introduce module overlaps.

The cost function is a generalization of the one used in the K-G-V algorithm (Section 2.2.1) together with a penalty function for module overlaps. To define the cost function, a set of vertical and horizontal grid lines are drawn across the chip. For each grid line l, let n_l be the number of nets crossing this line. Also, for each module A intersecting l, let I_A be the length of the intersection (either the height or the width of A). Let x_l be the sum of the lengths of the intersections of all the modules that intersect l. Clearly, the height (width) of the chip must be at least x_l if l is a vertical (horizontal) grid line. The first part of our cost function is the wiring cost

$$\mathbf{W} = \sum_{l} (n_l + \beta \cdot x_l)^2$$

where β is a constant. The terms x_l's (the modules' height/width) in the wiring cost is needed to reduce placement congestion. The value of β should be chosen such that contribution of these terms to the cost function is less than that of the n_l's. The second part of the cost function is the penalty function for module overlaps. The penalty term $O(i,j)$ for the overlap between module i and j is essentially the area of overlap but with the interiors of modules weighted more. (Note that by doing so, a small module on top of a large module will tend to be moved towards the edges rather than towards the center.) The penalty cost for module overlaps is

$$\mathbf{O} = \sum_{i \neq j} O(i,j).$$

The cost function is

$$\Psi = \mathbf{W} + \alpha \cdot \mathbf{O}$$

where α is a constant.

A problem of about 100 modules from an actual IBM chip design was used in [JeGe83] to illustrate the algorithm.

2.4.2. TimberWolf

The macro/custom–cell placement algorithm in TimberWolf is, in principle, similar to the Jespen and Gelatt's algorithm. However, TimberWolf has some additional features. One of them is to allow certain flexibility in the shape of the modules. Each module is allowed to have its aspect ratio chosen within a given range. A good placement algorithm should take advantage of such flexibility to select appropriate shapes for the modules in order to minimize the total chip area. Another new feature is to allow some pin terminals to have their positions on a module *uncommitted*. In some sense, TimberWolf has partially integrated the pin–assignment task with the placement task. Each uncommitted pin may be assigned to any one of a certain number of possible locations. Since the number of possible locations for an uncommitted pin might be very large (*e.g.* thousands), for execution time and storage considerations, it is assumed that there are only a limited number of pin *sites*, each representing a set of contiguous pin locations, evenly spaced on the boundary of a module. The *capacity* of a pin site is the total number of possible pin locations encompassed by the site. Uncommitted pins are assigned to pin sites instead of exact pin locations. Clearly, the total number of pins assigned to a pin site should not exceed its capacity. In the final placement, all pins assigned to a pin site will be assigned to the pin locations represented by the pin site.

In addition to the three types of moves employed in the previous algorithm, there are two additional moves:

M4. Aspect ratio change for a "flexible" module.
M5. Pin–site assignment change for an uncommitted pin or a group of uncommitted pins.

As in the TimberWolf standard–cell placement program, range limiters are used to control the scope of the moves.

The cost function consists of three parts. The first part is the typical estimation of the total wire length by summing up the half–perimeters of the bounding rectangles of all the nets. The second part is a penalty function for module overlaps based on the sum of the squares of the areas of overlap for all pairs of modules. The last part is a penalty function for capacity overflows at the pin sites.

The program was used to place a Honeywell circuit of 613 modules, each of which has 2 to 64 pins. TimberWolf used 18 CPU hours on a VAX 780 computer to place the modules. The result was compared with a manual placement obtained by a design team using 4 months. The estimated total wire length of the placement by TimberWolf was 21% less than that of the manual placement.

2.5. Other Stochastic Algorithms

In this section, we introduce two placement algorithms which are based on other stochastic search techniques. The first one is a gate-array placement algorithm called GINIE developed by Cohoon and Paris [CoPa86]. This algorithm is based on the genetic algorithm approach [Holl75]. The second one is the placement algorithm in the ESP standard-cell placement package developed by Kling and Banerjee [KlBa87]. This algorithm is based on the idea of simulated evolution. Both the genetic algorithm approach and the simulated evolution approach are based on the general idea of modelling an optimization problem as a natural evolutionary process in biological systems. The performances of these two placement algorithms, as reported in [CoPa86] and [KlBa87], are comparable with that of simulated-annealing based algorithms. In the rest of this section, we shall describe the placement algorithms GINIE and ESP in more detail.

2.5.1. Genetic Placement

Genetic Algorithm

Genetic algorithm is a general methodology for searching a solution space based on an analogy to biological evolution. As an optimization technique, such a methodology produces good solutions by examining and manipulating simultaneously a set of possible solutions. Let Π be the solution space. Each solution in Π is represented by a string of symbols called *allele*. A genetic algorithm produces a sequence of identical sized sets P_0, P_1, P_2, P_3, ..., where each P_i is a subset of Π. The set of solutions P_i is referred to as the *population* in the *i th generation*. The population "evolves" from one generation to the next by means of the *crossover* and *mutation* operations. The *crossover operator* σ is a function from $\Pi \times \Pi$ to Π. The *mutation operator* μ is a function from Π to Π. For every pair of solutions x and y, a new solution $\sigma(x,y)$ is created by combining x and y in such a way that they both contribute to the information carried by the solution $\sigma(x,y)$. (The solutions x and y are the *parents* and $\sigma(x,y)$ is the *offspring*.) At each generation, a fraction of the population is chosen as parents. The crossover operator σ is then applied to these parents to create a set of offspring. A set of solutions of the same size as the original population is then selected from the combined set of the original population and the offspring as "survivors" in the next generation. Each such surviving solution may be subjected to mutation by the mutation operator μ. Mutation prevents loss of diversity within a population by introducing new solutions. As the evolutionary process proceeds from one generation to the next, better solutions (solutions with lower cost) tend to survive and "less fit" solutions tend to die off. The quality of the solutions in general improves from one generation to the next. Eventually, one or more optimal or near-optimal solutions emerge.

GINIE: *A Genetic Placement Algorithm*

GINIE is a gate-array placement algorithm based on the genetic algorithm approach. The cost function used in GINIE is similar to that used in the K-G-V algorithm (Section 2.2.1). Each placement solution is represented by a string where the i *th* allele contains the module in the i *th* position of the chip in row-major order. The initial population is constructed by two *population constructors* $\mathbf{E}_1$ and $\mathbf{E}_2$. Constructor $\mathbf{E}_1$ randomly places modules on the chip, whereas constructor $\mathbf{E}_2$ chooses modules to be placed in a net-by-net manner and places the chosen modules on the chip in row-major fashion. Initially, a net is selected at random and all of its modules are placed on the chip. The next net to be processed will be an unplaced net containing a module that was placed as recently as possible. If there is no such unplaced net, an arbitrary unplaced net will be selected. All unplaced modules for the selected net are then placed on the chip. This process is continued until all modules are placed. Experimental results suggested the rule of thumb of starting with a mixed initial population with 75% of the placements constructed by $\mathbf{E}_2$ and the remaining 25% by $\mathbf{E}_1$.

The parents are chosen probabilistically according to their costs. Only those solutions with above-average costs are considered. The crossover operator σ performs a "cut-and-paste" type of operation on the two parents. Let x and y be the parents. Let z be an identical copy of y. A $j \times j$ sub-section C of x is randomly selected, where j has a truncated normal distribution with mean 3 and variance 1. Let D be the sub-section of modules in z that occupy the locations corresponding to C's locations. Let $C-D$ be the modules in C but not in D. Let $D-C$ be the modules in D but not in C. For the placement z, each module in $D-C$ is moved to a slot currently occupied by a module in $C-D$. Once all such modules are moved, C is copied into D. The "modified" z is defined to be the offspring $\sigma(x,y)$.

Let O_i be the set of offsprings of the population P_i. Let Q denote $P_i \cup O_i$. There are two ways to select a set of solutions of the same size as P_i from Q. The first way is to choose solutions randomly from Q. This method is more suitable for highly structured problem. The second way is make the selection according to a certain probability distribution favoring solutions with lower cost. This method is more suitable for less structured problem. The population P_{i+1} of the next generation is formed by applying the mutation operator μ to the solutions in Q. For the mutation operator μ, a net ξ is selected to be rearranged. Let m_s be a module in the net ξ chosen randomly. Let m_t be a module in the net ξ that is farthest away from m_s. A suitable location l adjacent to m_s is then determined. All modules along a Manhattan path from l to m_t are then shifted by one position. Finally, m_t is assigned to location l.

The algorithm GINIE terminates when there is no improvement in the best solution for 10,000 generations. The population size is set to 50, and 12 offsprings are created for each generation. GINIE was compared with a simulated-annealing based placement algorithm. For the five circuits tested, the results obtained by GINIE were comparable to those obtained by simulated annealing. Even though the qualities of the placements were not significantly different, in three of the five cases GINIE produced slightly better results. The running time was about the same for both algorithms.

2.5.2. Simulated Evolution Placement

We now describe a standard-cell placement algorithm called ESP [KlBa87]. The presentation here is quite brief and we refer the readers to [KlBa87] for many of the implementation details. The ESP placement algorithm is based on a heuristic method of simulating the natural selection process in biological environments. A similar concept was previously used in the area of logic minimization [Dunh63]. The objective of the ESP placement algorithm is to minimize the total wire length. As in other placement algorithms, the total wire length of each net is estimated by the half–

perimeter of the bounding rectangle of the net. In order to measure the quality of the placement of the modules of a net, we need to compare the estimated wire length of the net with its optimal value. An estimate of the optimal wire length of a net can be computed by assuming all the modules in the net are closely placed next to each other. The *placement value* of a net is the ratio of its estimated optimal wire length to its current estimated wire length in the placement. The *goodness value* of a module is computed by first taking the average of the placement values of all the nets it is connected to and then normalizing the result to a value between 0 and 100(%). The goodness value of a module represents the likelihood that the module is in the "right" place.

ESP starts with an initial solution z_0, and iteratively produces a sequence of placement solutions z_1, z_2, z_3, ... Each solution z_i "evolves" to the next solution z_{i+1} in two steps. The first step is either a *mutation* or to leave the solution unchanged. A *mutation* is a sequence of arbitrary pairwise exchanges of modules. The number of exchanges in a mutation is controlled by a user specified parameter. In ESP, mutation takes place only occasionally. (There is a procedure to remove overlaps or empty spaces, if any, created by mutation.) After the first step, the second step is to remove some of the modules from the current solution and re-allocate them to new locations. The set of modules that will remain in its present position is determined by comparing, for each module, the goodness value of the module and a random number between 0 and 100(%) generated for the module. If the goodness value is greater than the random number, the module will survive in its current position. Otherwise, it is removed and placed in a queue for re-allocation. The modules in the queue are then sorted according to the number of nets connected to each module. The idea is to put those modules with the most connections at the head of the queue. After that, the first module is removed from the queue and tentatively placed at all empty locations. An estimated goodness value of the module is efficiently computed for each trial location. The module is then placed at the location that maximizes its goodness value. The process is repeated for the other modules in the queue until all modules are placed. After this step, the resulting solution is z_{i+1}. (Again, there is a procedure to remove overlaps or empty spaces, if any, created by the re-allocation of the modules.)

Note that evolution of one solution to the next one might amount to a fairly global change. In fact, the number of modules scheduled for re-allocation can be as much as half of the total number of modules at the begining, but it slowly decreases to only a few near the end.

The algorithm terminates when no further improvement in total wire length can be achieved or when a user specified parameter such as CPU time of total wire length is reached. ESP was compared with TimberWolf. For the five circuits tested, experimental results with that obtained by ESP compared favorably with that obtained by TimberWolf. The results (in terms of the estimated total wire length) obtained by ESP were at best 15% less than that of TimberWolf and in the worst case 8% larger. The running time of ESP was in all cases significantly less than that of TimberWolf.

2.6. Concluding Remarks

Solution of the placement problem is one of the earlier applications of the method of simulated annealing. Many good simulated annealing based placement algorithms have been developed since its introduction. The method of simulated annealing appears to be particularly successful in solving various types of placement problems. In this chapter, we presented algorithms for gate-array placement, standard-cell placement, and macro/custom cell placement. For the gate-array placement problem, we presented the placement algorithm proposed in [KiGV83] by Kirpatrick, Gelatt, and Vecchi, and the gate-array placement algorithm in the TimberWolf placement and routing package [SeSa85]. For the standard-cell placement problem, we presented the standard-cell placement algorithm in TimberWolf [SeSa85, SeSa86] and the algorithm proposed by Grover in [Grov86] and [Grov87]. For the macro/placement problem, we presented the algorithm proposed by Jespen and Gelatt in [JeGe83] and the macro/custom-cell placement algorithm in the TimerWolf package. Finally, we also presented two placement algorithms that are based on similar stochastic techniques. The first is a gate-array placement algorithm called GINIE developed by Cohoon and Paris [CoPa86]. The second is the standard-cell placement algorithm proposed in [KlBa87] by Kling and Banerjee. Both algorithms are based on the general idea of modelling an optimization problem as a natural evolutionary process in biological systems. All the algorithms presented in this chapter produce good quality placement solutions.

CHAPTER 3

FLOORPLAN DESIGN

3.1. Introduction

Floorplan design [MaMH82, Otte82, LaDi85, PrCh85, WoWT86, WoLi86, WoLi87a] is the first stage of VLSI circuit layout. It is the problem of allocating spaces to a given set of circuit modules in the plane to minimize a weighted sum of the following two quantities: (1) the area of the bounding rectangle containing all the modules; and (2) an estimation of the total interconnection wire length (or any suitable proximity measure). A given module can be classified as *rigid* or *flexible*. A module is said to be ***rigid*** if its shape and dimensions are fixed. Pre-designed library macro-cells are examples of rigid modules. In this chapter, we consider both rectangular and L-shaped rigid modules. (A programmable logic array is an example of an L-shaped module.) A module is said to be *flexible* if its shape and dimensions are not fixed. Such flexibility represents the designer's freedom to manipulate the modules' internal structure at the floorplanning stage of design. We assume all flexible modules are rectangular. For each flexible module, we are given its area and limits of its allowed aspect ratio, where aspect ratio equals height divided by width. Floorplan design is a generalization of the classical placement problem in which all modules are rectangular rigid modules. Traditional placement algorithms are no longer effective for this more general problem. For each flexible module, a floorplan design algorithm should be able to take advantage of the freedom of selecting a representative among many alternatives. For each L-shaped module, a floorplan design algorithm should consider the specific shape of the module rather than just replacing the module by a bounding rectangle (which in general introduces unnecessary dead spaces).

A *floorplan* for n given modules (named 1,2,...,n) consists of an enveloping rectangle R subdivided by horizontal and vertical line segments into n or more non–overlapping rectilinear regions; n of these regions are labeled 1,2,...,n. Region i must be large enough to accommodate module i. (Unlabeled regions, if any, correspond to dead spaces that sometimes are needed to achieve a more compact packing of the modules.) We require that the aspect ratio of R be between two given numbers p and q. We are also given an $n \times n$ interconnection matrix $C = (c_{ij})_{n \times n}$, with $c_{ij} \geq 0$, $1 \leq i,j \leq n$, which provides information on the wiring density between each pair of modules. The *center* of a region is the center of mass of the region. The *distance* between two regions is the Manhattan distance between their centers. For every pair of modules i and j, let d_{ij} be the distance between regions i and j. We use $W = \sum_{1 \leq i,j \leq n} c_{ij} d_{ij}$ as an estimate of the total interconnection wire length. Let A be the area of R. A and W will be referred to as *area* and *total wire length* of the floorplan, respectively. We use $A + \lambda W$ to measure the quality of a floorplan, where λ is a user specified constant that controls the relative importance of A and W. The objective of the floorplan design problem is to find a floorplan such that $A + \lambda W$ is minimized. (The algorithms in this chapter are quite flexible with respect to different quality measures. In particular, W can be replaced by any other proximity measure.)

In this chapter, we present two algorithms [WoLi86, WoLi87a] for the floorplan design problem. These two algorithms are similar in spirit. They both use Polish expressions to represent floorplans and employ the method of simulated annealing. The first algorithm [WoLi86] is designed for the case where all modules are rectangular. The second algorithm [WoLi87a] is designed for the case where the modules are either rectangular or L–shaped. Many other floorplanning algorithms derive the final solution in two stages. They first determine the relative positions of the modules using primarily interconnection information, then they use the area and shape information to minimize the area of the bounding rectangle, whereas the algorithms in this chapter consider simultaneously the interconnection information as well as the area and shape information. Also, other algorithms that employ the method of simulated annealing either use a representation that leads to a larger number of states and thus ultimately a slower rate of convergence [SeSa85], or apply the search method only at a particular stage of a heuristic floorplan design algorithm [OtGi84]. Finally, most algorithms can only handle rectangular modules.

This chapter consists of two parts. Part 1 is for the case where all modules are rectangular. Part 2 is for the case where the modules are either rectangular or L–shaped.

3.2. Part 1: Rectangular Modules

In this section, we assume the given modules are all rectangular. A *rectangular floorplan* is a floorplan where all the regions are rectangles. The rectangular regions in a rectangular floorplan are referred to as *basic rectangles*. We shall only consider rectangular floorplans. Let (A_1,r_1,s_1), (A_2,r_2,s_2), $\cdots$, (A_n,r_n,s_n) be a list of n triplets of numbers corresponding to the n given modules. The triplet of numbers (A_i,r_i,s_i), with $r_i \le s_i$, specifies the area and the limits of the allowed aspect ratio for module i. Each module can also have a *fixed* or *free* orientation. If the orientation is free, rotation of the module (by 90°) is allowed. Otherwise, rotation of the module is not allowed. Let O_1 be the set of modules with fixed orientation, and O_2 be the set of modules with free orientation. Let w_i be the width and h_i be the height of module i, we must have :

(1) $w_i h_i = A_i$

(2) $r_i \le h_i/w_i \le s_i$ if $i \in O_1$

(3) $r_i \le h_i/w_i \le s_i$ or $1/s_i \le h_i/w_i \le 1/r_i$ if $i \in O_2$.

Note that module i is a rigid module iff $r_i = s_i$. Let x_i be the width and y_i be the height of region i. Clearly, we must have $x_i \ge w_i$ and $y_i \ge h_i$.

3.2.1. Slicing Floorplans

To *cut* a rectangle, we mean to divide the rectangle into two rectangles by either a vertical or a horizontal line. A *slicing floorplan* is a rectangular floorplan with n basic rectangles that can be obtained by recursively cutting a rectangle into smaller rectangles. (See Fig.3.1(a).) A slicing floorplan can be represented by an oriented rooted binary tree, called a *slicing tree*. (See Fig.3.1(b).) Each internal node of the tree is labeled either $*$ or $+$,

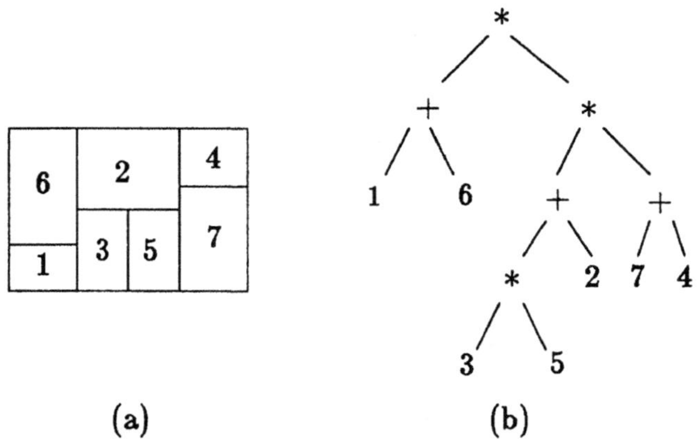

Fig. 3.1 Slicing floorplan and its slicing tree representation.

corresponding to either a vertical or a horizontal cut, respectively. Each leaf corresponds to a basic rectangle and is labeled by a number between 1 and n.

A slicing tree is a hierarchical description of the types of the cuts (vertical or horizontal) in a slicing floorplan. However, no dimensional information on the position of each cut is specified. In general, there are many floorplans represented by the same slicing tree. These floorplans differ in the geometric dimensions for the basic rectangles as well as in the adjacency relationship among the basic rectangles. We now define an ***equivalence relation*** $\sim$ on the set of all slicing floorplans. Let A and B be two slicing floorplans. We define $A \sim B$ iff they have the same slicing tree representation. The equivalence relation $\sim$ partitions the set of slicing floorplans into equivalence classes. Each equivalence class of slicing floorplans with n basic rectangles is called a ***slicing structure***.

Note also that for a given slicing floorplan, there may be more than one slicing–tree representation. (See Fig.3.2.) The different slicing trees that represent the same floorplan correspond to different orders in which consecutive vertical cuts and consecutive horizontal cuts are made. A ***skewed slicing tree*** is a slicing tree in which no node and its right son have the same label in $\{*,+\}$. (See Fig.3.3.) A slicing tree obtained by making consecutive vertical cuts in the order of from right to left, and making consecutive horizontal cuts in the order of from top to bottom is a skewed slicing

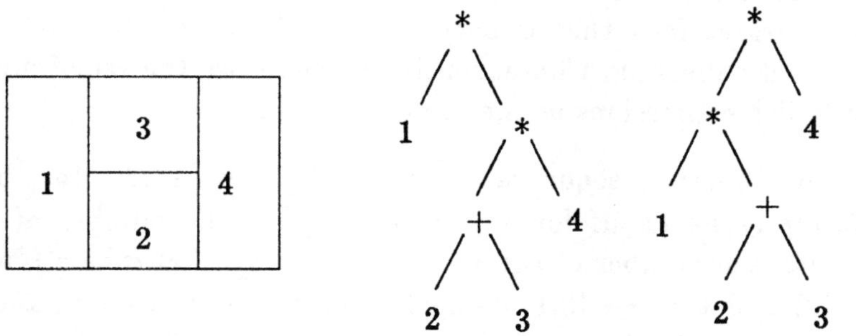

Fig. 3.2 Two different slicing trees for the same slicing floorplan.

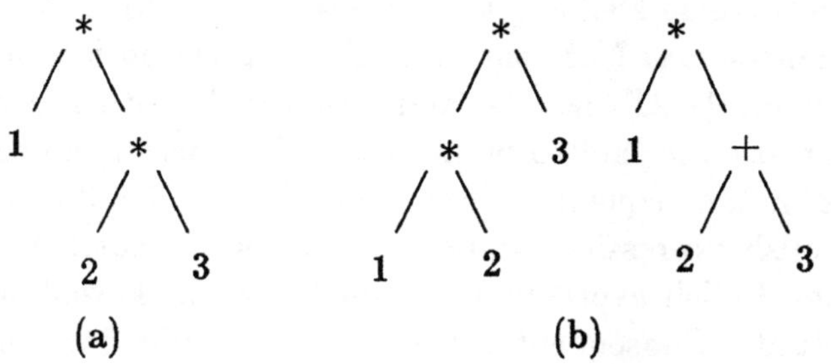

Fig. 3.3 Skewed and non–skewed slicing trees. (a) A non–skewed tree. (b) Skewed trees.

tree. All other slicing trees representing the same floorplan are not skewed. Hence we can use skewed slicing trees to represent slicing floorplans. We have the following lemma, the proof of which is rather obvious and is thus omitted.

Lemma 3.1. There is a 1–1 correspondence between all skewed slicing trees with n leaves and all slicing structures with n basic rectangles.

3.2.2. Solution Space

A slicing structure represents a set of equivalent slicing floorplans. We shall see in Section 3.2.4 that we can efficiently select the "best" floorplan among all these equivalent floorplans. Thus, instead of using the set of all slicing floorplans as the solution space, we can use the set of all slicing structures as the solution space. This substantially reduces the size of the solution space. We shall use a representation of slicing structures called *normalized Polish expressions* that is particularly suitable for the method of simulated annealing. Consequently, we shall use the set of normalized Polish expressions as the solution space.

A binary sequence $b_1 b_2 \cdots b_m$ is said to be a *balloting sequence* iff for any k, $1 \leq k \leq m$, the number of 0's is less than the number of the 1's in $b_1 b_2 \cdots b_k$. Let σ be a function $\sigma : \{1,2,...,n,*,+\} \rightarrow \{0,1\}$ defined by $\sigma(i) = 1$, $1 \leq i \leq n$, and $\sigma(*) = \sigma(+) = 0$. A sequence $\alpha_1 \alpha_2 \cdots \alpha_{2n-1}$ of elements from $\{1,2,...,n,*,+\}$ is said to have the *balloting property* if $\sigma(\alpha_1)\sigma(\alpha_2) \cdots \sigma(\alpha_{2n-1})$ is a balloting sequence.

If we traverse a slicing tree in postorder [AhHU74], we obtain a *Polish expression*. Note that a Polish expression is a sequence of elements from $\{1,2,...,n,*,+\}$ and has the balloting property. A Polish expression is said to be *normalized* if there is no consecutive $*$'s or $+$'s in the sequence. (For example, $1\ 2 + 4\ 3 * +$ is a normalized Polish expression, while $1\ 2 + 4\ 3 * *$ is not.) It is easy to see that the Polish expression obtained from a skewed slicing tree is normalized. Consequently we have the following lemma and theorem. (Theorem 3.1 follows from Lemma 3.1 and Lemma 3.2.)

Lemma 3.2. There is a 1–1 correspondence between the set of normalized Polished expressions of length $2n-1$ and the set of skewed slicing trees with n leaves.

Theorem 3.1. There is a 1–1 correspondence between the set of normalized Polish expressions of length $2n-1$ and the set of slicing structures with n basic rectangles.

A slicing tree can be viewed either as a *top down* or a *bottom up* description of a slicing floorplan. From a top down point of view, a slicing tree specifies how a given rectangle is cut into smaller rectangles by horizontal and vertical cuts. From a bottom up point of view, a slicing floorplan can also be described by how smaller slicing floorplans are combined recursively to yield larger slicing floorplans. Indeed, we can interpret the symbols $*$ and $+$ as two binary operators between slicing floorplans. If A and B are slicing floorplans, we can interpret $A+B$ and $A*B$ as the resulting slicing floorplans obtained by placing B on top of A, and B to the right of A, respectively, as shown in Fig.3.4. (From now on, we shall refer to the elements in $\{1,2,...,n\}$ as *operands*, and the elements in $\{*,+\}$ as *operators*.)

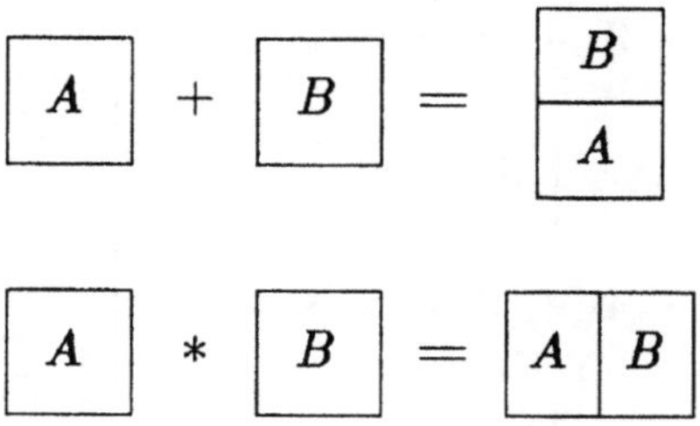

Fig. 3.4 Binary operators for slicing floorplans.

3.2.3. Neighboring Solutions

A sequence $d_1 d_2 \cdots d_k$ of k operators is called a *chain* of *length* k. (Note that $d_i \neq d_{i+1}$ in a normalized Polish expression, for all $1 \leq i \leq k-1$.) We shall use $l(c)$ to denote the length of a chain c. A chain of length 0 is defined to be the empty sequence. It is clear that for every $k > 0$, there are only two possible types of chains of length k in a normalized Polish expression: $*+*+* \cdots$ and $+*+*+ \cdots$. We define the *complement* of a chain to be the chain obtained by interchanging the operators $*$ and $+$ (*e.g.* the complement of $*+*++*$ is $+*+**+$). Let $\boldsymbol{\alpha} = \alpha_1 \alpha_2 \cdots \alpha_{2n-1}$ be a normalized Polish expression. Note that $\boldsymbol{\alpha}$ can also be written as $c_0 \pi_1 c_1 \pi_2 c_2 \cdots c_{n-1} \pi_n c_n$, where $\pi_1 \pi_2 \cdots \pi_n$ is a permutation of $1,2,...,n$, the c_i's are chains (possibly of zero length), and $\sum_i l(c_i) = n-1$. (For example, if $\boldsymbol{\alpha} = 1\,2\,3*+5\,4+*$, we have $\pi_1\pi_2\pi_3\pi_4\pi_5 = 1\,2\,3\,5\,4$, $c_2 = *+$, $c_5 = +*$, and $c_0 = c_1 = c_2 = c_4 =$ the empty sequence.) Two operands in $\boldsymbol{\alpha}$ are said to be *adjacent* iff they are consecutive elements in $\pi_1 \cdots \pi_n$. An operand and an operator are said to be *adjacent* iff they are consecutive elements in $\alpha_1 \alpha_2 \cdots \alpha_{2n-1}$. We define three types of moves M1, M2 and M3 that can be used to modify a given normalized Polish expression. The definitions of M1, M2 and M3 are as follows:

M1. Swap two adjacent operands.
M2. Complement a chain of nonzero length.
M3. Swap two adjacent operand and operator.

See Fig.3.5 for a pictorial demonstration of these three types of moves.

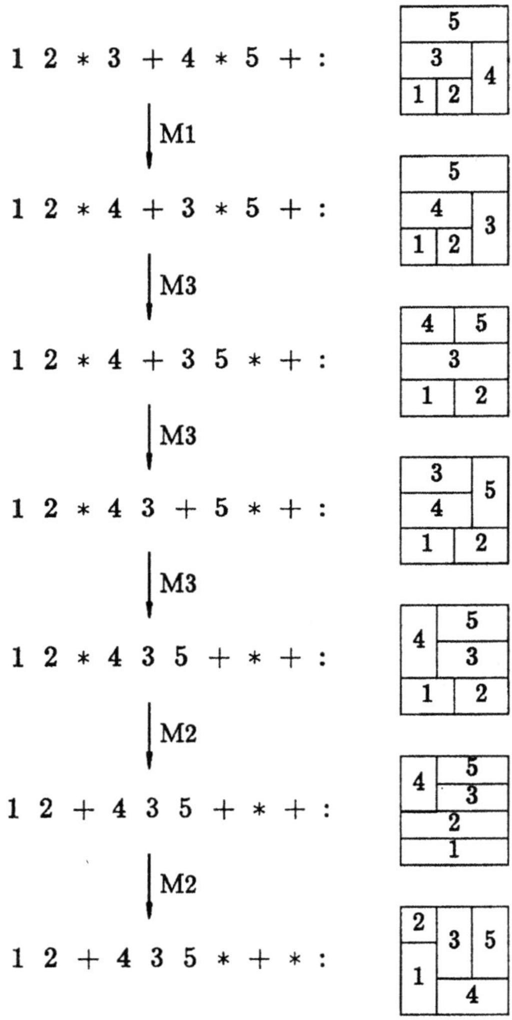

Fig. 3.5 Illustration of the moves.

It is clear that M1 and M2 always produce a normalized Polish expression. This is not the case for M3. In fact, M3 might produce a sequence that contains identical consecutive operators or a sequence that violates the balloting property. We shall only accept those M3 moves that result in normalized Polish expressions.

(Whether a given M3 move will result in a normalized Polish expression can be tested efficiently in constant time.)

Two normalized Polish expressions are said to be *neighbors* if one can be obtained from the other via one of these three moves. For a given normalized Polish expression α, we first randomly select a type of move, and then randomly select a neighbor according to the type of move selected. The three types of moves are sufficient to ensure that it is possible to transform any normalized Polish expression into any other via a sequence of moves as the following theorem shows.

Theorem 3.2. Let α and α' be two normalized Polish expressions. There exists a sequence of at most $2n^2 + 2n$ neighboring normalized Polish expressions between α and α'.

Proof. We observe that given any normalized Polish expression α, α is reachable from $\alpha_0 = 12{*}3{*} \cdots {*}n$ via a sequence of moves, and vice versa. Therefore, for any pair of normalized Polish expressions α and α', there exists a sequence of moves that will transform α' into α with α_0 as an intermediate expression. In order to go from α to α_0, we need at most $n^2/2$ moves of type M1 to sort the operands into the ordering 1, 2, 3, ..., n; at most another $n^2/2$ moves of type M3 to convert the expression into the form $12{\odot}3{\odot} \cdots {\odot}n$ where $\odot$ is an operator; and at most n moves of type M2 to convert all operators into $*$. Hence it takes at most $n^2 + n$ moves to convert α into α_0. Similarly, it takes at most $n^2 + n$ moves to convert α_0 to α'. Therefore, the total number of moves needed is at most $2n^2 + 2n$. □

3.2.4. Cost Function

Let α be a normalized Polish expression. The expression α represents a set $\mathbf{S}_\alpha$ of equivalent slicing floorplans. Roughly speaking, the relative positions of the modules in equivalent floorplans are essentially the same. Intuitively, those floorplans that have smaller area will in general also have shorter total wire length because the modules are pulled "closer together" in these floorplans. Let $\mathbf{f}_\alpha$ be a floorplan in $\mathbf{S}_\alpha$ with minimum area. Let $A(\alpha)$ and $W(\alpha)$ be the area and the total wire length of $\mathbf{f}_\alpha$, respectively. The cost function is of the form $\Psi(\alpha) = A(\alpha) + \lambda\, W(\alpha)$ where λ is a constant. We show in this section how to compute efficiently $\Psi(\boldsymbol{\alpha})$ for a given normalized Polish expression $\boldsymbol{\alpha}$. The method of cost computation is based on [Otte83] and [Stoc83]. Since in the simulated annealing algorithm each move only leads to a minor modification of the present Polish expression, we can take advantage of the cost computation for the preceding move to substantially save computational effort for the current cost computation.

3.2.4.1. Basic Method

Shape Curve

Let Γ be a continuous curve on the plane. Γ is said to be *decreasing* if for any two points (x_1,y_1), (x_2,y_2) on Γ with $x_1 \leq x_2$, we must have $y_1 \geq y_2$. Γ is a *shape curve* if it satisfies the following conditions: (1) it is decreasing and lies completely in the first quadrant; (2) $\exists\ k > 0$ such that all lines of the form $x = a$, $a > k$, intersect Γ; and (3) $\exists\ k > 0$ such that all lines of the form $y = b$, $b > k$, intersect Γ. Note that a shape curve Γ partitions the first quadrant into two connected regions. One of the connected regions that contains the points (a,a) for all large a's is called the *feasible region*. (See Fig.3.6.)

The shape curves in Fig.3.7 correspond to different kinds of shape constraints where the shaded areas are the feasible regions. A point in the feasible region corresponds to the dimensions of a basic

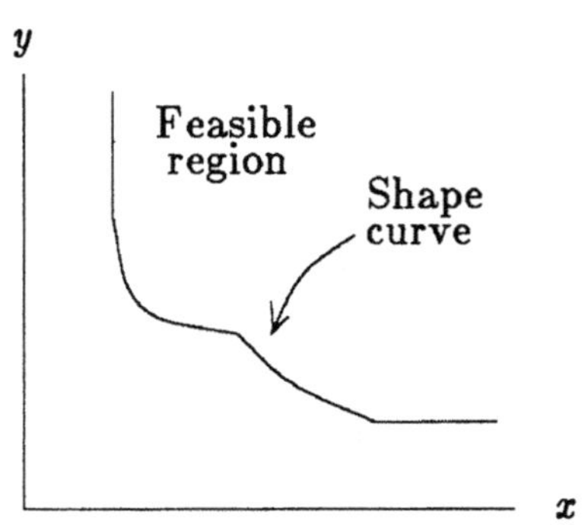

Fig. 3.6 Shape curve and feasible region.

rectangle that can accommodate a given module. (The x and y coordinates of the point are the horizontal and vertical dimensions of the basic rectangle.) Fig.3.7(a) and Fig.3.7(b) correspond to the case in which the module is rigid. Fig.3.7(c) and Fig.3.7(d) correspond to the case in which the module is flexible. Let H be the hyperbola $xy = A_i$, L_1 be the line $y = s_i x$, L_2 be the line $y = r_i x$, L_3 be the line $y = (1/r_i)x$, and L_4 be the line $y = (1/s_i)x$. In Fig.3.7(a),(b),(c),(d), a, b, c and d are the intersections between the hyperbola H and the lines L_1, L_2, L_3 and L_4, respectively.

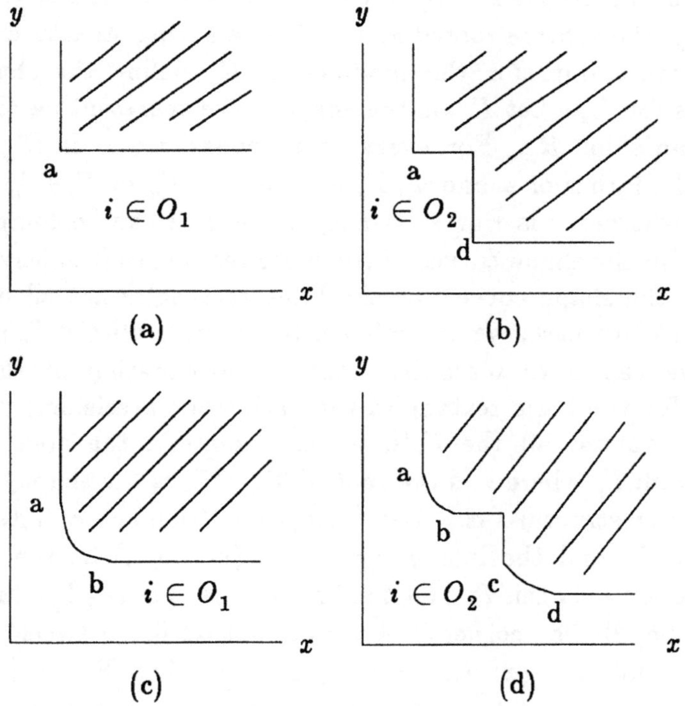

Fig. 3.7 Shape curves for different shape constraints.

Let Γ and Λ be two shape curves. We define $\Gamma + \Lambda$ to be the curve $\{(u,v+w) \mid (u,v) \in \Gamma \text{ and } (u,w) \in \Lambda\}$ and define $\Gamma * \Lambda$ to be the curve $\{(u+v,w) \mid (u,w) \in \Gamma \text{ and } (v,w) \in \Lambda\}$. In other words, $\Gamma + \Lambda$ is obtained by adding the two curves Γ and Λ along the $y-$direction. $\Gamma * \Lambda$ is obtained by adding the two curves Γ and Λ along the $x-$direction. It is easy to see that $\Gamma + \Lambda$ and $\Gamma * \Lambda$ are also shape curves. Moreover, they are piecewise linear if Γ and Λ

along the x-direction. It is easy to see that $\Gamma + \Lambda$ and $\Gamma * \Lambda$ are also shape curves. Moreover, they are piecewise linear if Γ and Λ are both piecewise linear. For piecewise linear shape curves, there is an efficient algorithm to compute $\Gamma + \Lambda$ and $\Gamma * \Lambda$. Since a piecewise linear shape curve is completely characterized by an ordered list of all the "corners" of the curve, hence to add two piecewise linear shape curves (along either directions), we only need to add up the curves at the "corners."

Area Computation

Let $\mathbf{T}_\alpha$ be the slicing tree corresponding to $\boldsymbol{\alpha}$. For each node v in $\mathbf{T}_\alpha$, the subtree rooted at v defines a slicing structure R_v. The shape constraints for the modules in R_v define the shape constraints for R_v. Let Γ_v be the shape curve representing the shape constraints for R_v. For every three nodes u,v,w in $\mathbf{T}_\alpha$ with v being the father of u and w, Γ_v is either $\Gamma_u * \Gamma_w$ or $\Gamma_u + \Gamma_w$ depending on whether v is $*$ or $+$. Hence, all the Γ_v's can be computed by adding up the shape curves of the basic rectangles (the leaves). We assume the shape curves for the basic rectangles are all piecewise linear. In this case, we can efficiently compute all the Γ_v's. (Note that we can have piecewise linear approximation of the shape curves for the basic rectangles with arbitrary precision.) Once we have computed all the Γ_v's, we can compute the area measure $A(\boldsymbol{\alpha})$ from Γ_r where r is the root of $\mathbf{T}_\alpha$. (Γ_r is the shape curve for the slicing structure $\boldsymbol{\alpha}$.) Let (a_1,b_1) be the point of intersection between Γ_r and the line $y = px$. Let (a_{l+1},b_{l+1}) be the point of intersection between Γ_r and the line $y = qx$. Let (a_2,b_2), (a_3,b_3), ... ,(a_l,b_l) be all the "corners" of the curve that lies between the lines $y = px$ and $y = qx$. (*i.e.* $p \le b_j/a_j \le q$.) Let $\mathbf{P} = \{ (a_1,b_1), (a_2,b_2), \ldots, (a_{l+1},b_{l+1}) \}$. The next theorem shows that we can compute $A(\boldsymbol{\alpha})$ by only examining the points in $\mathbf{P}$.

Theorem 3.3. $A(\boldsymbol{\alpha}) = a_i b_i$ where (a_i,b_i) is a point in $\mathbf{P}$ such that $a_i b_i$ is minimum.

Proof. Let (a,b) be the dimensions of a minimum area floorplan. It is clear that the point (a,b) must be on Γ_r. Since if that is not the case, the intersection between Γ_r and the line joining (0,0) and (a,b) would give the dimensions of a floorplan with smaller area. To show that (a,b) is one of the (a_i,b_i)'s, we observe that along a given line segment I in the first quadrant, the function $f(x,y) = xy$ attains its minimum value at one of the two end points of I. □

Wire Length Computation

Since $\mathbf{T}_\alpha$ is a slicing tree representation of $\mathbf{f}_\alpha$, each node v in $\mathbf{T}_\alpha$ corresponds to a rectangle K_v in $\mathbf{f}_\alpha$. Let (x_v, y_v) be the dimensions of K_v. After we have computed (x_r, y_r) for K_r where r is the root of $\mathbf{T}_\alpha$, we can recursively compute the dimensions of all the basic rectangles as follows: Suppose that we have already determined (x_v, y_v) for some internal node v. Let u be the left son of v and w be the right son of v. Consider the case in which the operator at v is +. That is, we have K_v is formed by putting K_w directly on top of K_u. We have $x_u = x_w = x_v$; $y_u = y_v - y_w$; and $y_w = y'$ where y' is the y-coordinate of the point of intersection between the line $x = x_v$ and the curve Γ_w. (If there is more than one intersection, choose the one with the smallest y-coordinate.) Similarly, for the case in which the operator at v is $*$, we have $y_u = y_w = y_v$; $x_u = x_v - x_w$; and $x_w = x'$ where x' is the x-coordinate of the point of intersection between the line $y = y_v$ and the curve Γ_w.

Let (c_v^x, c_v^y) be the coordinates of the center of K_v. (We assume that the southwest corner of $\mathbf{f}_\alpha$ is placed at the origin.) Let (l_v^x, l_v^y) be the coordinates of the point of the southwest corner of K_v. Clearly, we have $c_v^x = l_v^x + \frac{1}{2}x_v$ and $c_v^y = \frac{1}{2}l_v^y$. Consequently, to compute the centers of the basic rectangles, it suffices to compute (l_v^x, l_v^y) for all v. Note that $(l_r^x, l_r^y) = (0,0)$. We can recursively compute all (l_v^x, l_v^y)'s as follows. Suppose we have already determined (l_v^x, l_v^y) for some v. Let u be the left son of v and w be the right son of v. We have $(l_u^x, l_u^y) = (l_v^x, l_v^y)$. If the operator at v is +, we have $l_w^x = l_v^x$ and $l_w^y = l_v^y + y_u$. If the operator at v is $*$, we have $l_w^x = l_v^x + x_u$ and $l_w^y = l_v^y$.

Note that the centers of the basic rectangles are just the (c_v^x, c_v^y)'s for all leaf nodes v. It follows that the d_{ij}'s (the distances between the basic rectangles) and consequently $W(\alpha)$ can be easily computed from the (c_v^x, c_v^y)'s.

3.2.4.2. Incremental Computation of Cost Function

For a given normalized Polish expression, the shape curves associating with the nodes of its slicing tree are needed in both the area and the wire length computation. Let $\boldsymbol{\alpha}' = \alpha_1'\alpha_2' \cdots \alpha_{2n-1}'$ be the Polish expression obtained from $\boldsymbol{\alpha} = \alpha_1\alpha_2 \cdots \alpha_{2n-1}$ after a move. For $1 \leq i \leq 2n-1$, let T_i and T_i' be the trees rooted at α_i and α_i', and let Γ_i and Γ_i' be the shape curves for α_i and α_i', respectively. In the simulated annealing algorithm, each move leads to only a minor modification of the Polish expression currently being examined. Thus, in general, $\Gamma_i = \Gamma_i'$ for many i's. Therefore, in computing the cost for $\boldsymbol{\alpha}'$, we need only to update those shape curves that are changed. Such an observation will substantially improve the efficiency of the algorithm because we need to compute the costs for a large number of solutions. We will show in Theorem 3.4 that the two sets of shape curves $\{\Gamma_1,\Gamma_2, \cdots ,\Gamma_{2n-1}\}$ and $\{\Gamma_1',\Gamma_2', \cdots ,\Gamma_{2n-1}'\}$ differ only at a set of vertices that lie along one or two paths in each of $\mathbf{T}_{\boldsymbol{\alpha}}$ and $\mathbf{T}_{\boldsymbol{\alpha}'}$.

For a move of type M1 or M3, we swap two elements α_i and α_j. We shall refer to the elements α_i and α_j as the *base elements* in $\boldsymbol{\alpha}$, and the elements α_i' and α_j' as the *base elements* in $\boldsymbol{\alpha}'$. For a move of type M2, we complement a chain $\alpha_k\alpha_{k+1}\alpha_{k+2} \cdots \alpha_l$. We shall refer to the elements α_k, α_{k+1}, ..., α_l as the *base elements* in $\boldsymbol{\alpha}$, and the elements α_k', α_{k+1}', ..., α_l' as the *base elements* in $\boldsymbol{\alpha}'$. The following lemma follows from the definition of base elements.

Lemma 3.3. (1) α_i is a base element iff α_i' is a base element. (2) If α_i is not a base element, then $\alpha_i = \alpha_i'$.

Let us *mark* all the base elements together with all their ancestors in both $\boldsymbol{\alpha}$ and $\boldsymbol{\alpha}'$. Let $\mathbf{M}$ be the set of marked elements in $\boldsymbol{\alpha}$ and $\mathbf{M}'$ be the set of marked elements in $\boldsymbol{\alpha}'$. The elements which are not in $\mathbf{M}$ or $\mathbf{M}'$ are said to be *unmarked*. We have the following lemmas:

Lemma 3.4. α_i is unmarked iff α_i' is unmarked.

Proof. Note that T_i corresponds to $\boldsymbol{\beta}_i = \alpha_j\alpha_{j+1} \cdots \alpha_i$ and T_i' corresponds to $\boldsymbol{\beta}_i' = \alpha_k'\alpha_{k+1}' \cdots \alpha_i'$ where j and k are the unique indices such that $\sigma(\alpha_j)\sigma(\alpha_{j+1}) \cdots \sigma(\alpha_i)$ and $\sigma(\alpha_k')\sigma(\alpha_{k+1}') \cdots \sigma(\alpha_i')$ are balloting sequences. Suppose α_i is unmarked. This means that

β_i does not contain any base elements. It follows from Lemma 3.3 that $\alpha'_m = \alpha_m$ for all $j \leq m \leq i$. Hence $\sigma(\alpha'_j)\sigma(\alpha'_{j+1}) \cdots \sigma(\alpha'_i)$ is a balloting sequence. Therefore, we must have $k = j$ and $\beta'_i = \beta_i$. Again, it follows from Lemma 3.3 that β'_i does not contain any base elements. Therefore α'_i is unmarked. Similarly, we can show the converse by interchanging the roles of $\boldsymbol{\alpha}$ and $\boldsymbol{\alpha}'$ in the above arguments □

Lemma 3.5. If α_i is unmarked (or equivalently α'_i is unmarked), then $\Gamma_i = \Gamma'_i$.

Proof. It follows from Lemma 3.4 that $T_i = T'_i$. Thus, we have $\Gamma_i = \Gamma'_i$. □

Lemma 3.6. **M** corresponds to one or two paths in $\mathbf{T}_{\boldsymbol{\alpha}}$; **M**$'$ corresponds to one or two paths in $\mathbf{T}_{\boldsymbol{\alpha}'}$.

Proof. For a move of type M1 or M3, there are only two base elements in $\boldsymbol{\alpha}$. **M** corresponds to the two paths from the two base elements in $\boldsymbol{\alpha}$ to the root. For a move of type M2, let α_k, α_{k+1}, α_{k+2}, ..., α_l be the base elements in $\boldsymbol{\alpha}$. In this case, α_i is the right son of α_{i+1}, for all $k \leq i \leq l-1$. **M** corresponds to the path from α_k to the root. Similarly, we can show that **M**$'$ corresponds to one or two paths in $\mathbf{T}_{\boldsymbol{\alpha}'}$. □

It follows from Lemma 3.5 that we need only to recompute those shape curves that correspond to unmarked elements after each move. The next theorem follows directly from the above lemmas.

Theorem 3.4. Let $\boldsymbol{\alpha}'$ be the Polish expression obtained from $\boldsymbol{\alpha}$ after a move. The shape curves Γ_i's and the shape curves Γ'_i's differ only at a set of vertices that lie along one or two paths in each of $\mathbf{T}_{\boldsymbol{\alpha}}$ and $\mathbf{T}_{\boldsymbol{\alpha}'}$.

Fig.3.8 illustrates the statement in Theorem 3.4. In the given example, the Polish expression is of the form $\cdots 17 + 16 \cdots * 6 \cdots 9 + * 8 \cdots$. Consider the type M3 move of swapping $*$ and 6. Fig.3.8(a) and Fig.3.8(b) show the path(s) corresponding to this move in $\mathbf{T}_{\boldsymbol{\alpha}}$ and $\mathbf{T}_{\boldsymbol{\alpha}'}$, respectively.

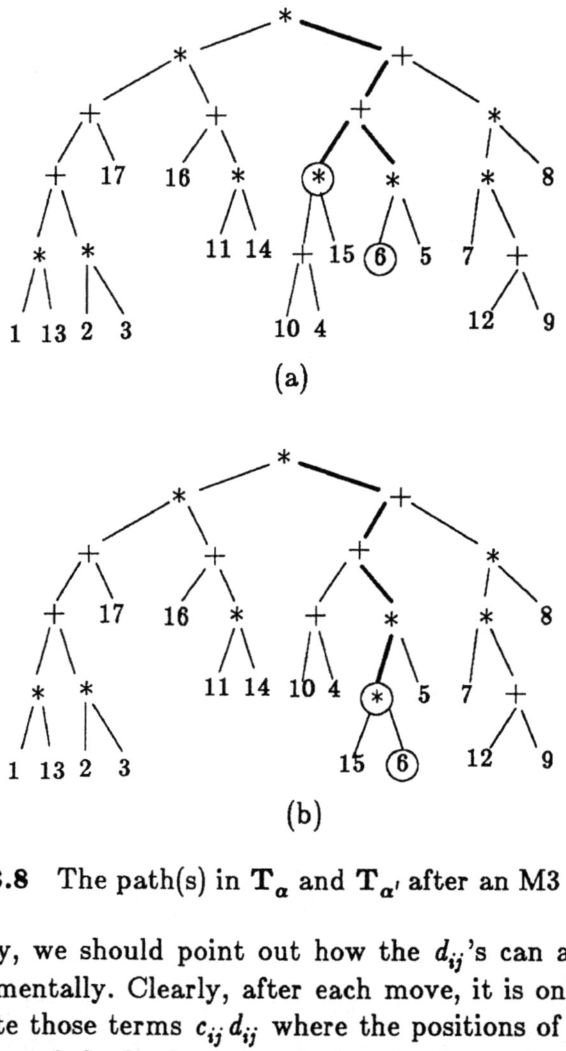

Fig. 3.8 The path(s) in $\mathbf{T}_\alpha$ and $\mathbf{T}_{\alpha'}$ after an M3 move.

Finally, we should point out how the d_{ij}'s can also be computed incrementally. Clearly, after each move, it is only necessary to recompute those terms $c_{ij} d_{ij}$ where the positions of one or both of the centers of the basic rectangle i and j have changed. Unfortunately, in general, even for a minor modification of a Polish expression, the positions of the centers of many of the basic rectangles will be changed (although most of the changes are small). In view of this, we can reduce the computation time by using a less accurate wire length estimation. Suppose that there is a grid of size ρ (*i.e.* the distance between every two adjacent lines in the grid is ρ) imposed on top of the floorplan. We define the *modified center* of a basic rectangle as the grid point that is closest to the center of the basic rectangle. (In the case when there are more than one such grid points, we use the one with the smallest x and y coordinates.) For every pair of basic rectangles i and j, we define d'_{ij} as the distance between their modified centers. Let $W'(\alpha) = \sum_{1 \leq i,j \leq n} c_{ij} d'_{ij}$. We can use $A(\alpha) + \lambda W'(\alpha)$ as the cost function. (Note that $W'(\alpha) = W(\alpha)$ if ρ is 0.) Clearly, the number of *modified centers* which remain unchanged after a move decreases as ρ increases. Consequently, the computation time for updating the d'_{ij}'s is inversely proportional to ρ.

3.2.5. Annealing Schedule

The temperature schedule is of the form $T_k = r \cdot T_{k-1}$, $k = 1, 2, 3, \ldots$. A typical value for r is 0.85. The initial temperature T_0 is determined by performing a sequence of random moves and computing the quantity Δ_{avg}, the average value of the magnitude of change in cost per move. We should have $e^{-\Delta_{avg}/T_0} = P \simeq 1$ so that there will be a reasonable probability of acceptance at high temperatures. This suggests that $T = -\Delta_{avg}/\ln(P)$ is a good choice for T_0.

The algorithm can start with any initial normalized Polish expression. The default initial Polish expression is $12{*}3{*}4{*} \cdots {*}n{*}$ which corresponds to placing the n modules horizontally next to each other. This Polish expression is usually far from the optimal solution.

At each temperature, enough moves are attempted until either there are N downhill moves or the total number of moves exceeds $2N$, where $N = \gamma n$ and γ is a user specified constant. The annealing process is terminated when the number of accepted moves is less than 5% of all moves made at a certain temperature or the temperature is low enough.

3.2.6. Experimental Results

The algorithm has been implemented in Pascal on a PYRAMID computer. The experimental results are summarized in Tables 3.1, 3.2 and 3.3. The modules in all the tests problems are flexible modules with free orientations allowed and $r_i = 1/s_i$. The

Problem	n	λ_w	Total Area	Random Feasible Solution		Initial S.A. Solution		Final S.A. Solution	
				A	W	A	W	A	W
P1	15	1	137.08	491.64	106.02	514.59	59.41	137.86	34.97
P2	20	1	198.88	745.35	258.84	885.71	202.3	202.17	80.49
P3	20	0	197.15	808.72	423.90	819.64	276.5	198.98	197.4
P4	25	0	244.68	1165.7	398.21	1334.9	296.6	245.43	209.7
P5	25	1	238.15	1026.6	576.71	876.04	344.3	244.63	151.9
P6	30	0.5	333.92	1549.6	1023.2	2458.2	865.7	340.15	294.9
P7	30	0	314.45	1476.9	1095.7	2130.1	936.2	319.40	429.2
P8	40	1	407.44	1934.2	1002.3	3965.4	999.3	422.95	265.8

Table 3.1

running time for the test problems range from 1 CPU minute for the 15 modules problem to 13 CPU minutes for the 40 modules problem.

The problems in Table 3.1 are randomly generated. The areas of the modules (A_i) are chosen uniformly between 1 and 20. For P4 and P5, all modules have the same aspect ratio (s_i). It is 2 for P4 and 3 for P5. For the other problems, each s_i is chosen uniformly between 1 and 3. The maximum aspect ratio allowed for the final chip is 2. The interconnection matrices are also randomly generated such that the weights (c_{ij}) are between 0 and 1. Various values of λ are used in these test problems. Total Area in the table is the sum of the areas of the given modules and hence is a lower bound on the area of the final chip. Comparing the results for problems with the same number of modules (*i.e.* P2 & P3, P4 & P5, and P6 & P7), we note that the final area (A) of P2 and P3 are both very close to optimal while the final total wire length (W) of P3 is substantially higher than that of P2. Since their interconnection matrices were generated using the same probability distribution, one might expect the values of W to be about the same. The large difference in the final values of W is due to the fact that the value of λ is 0 for P3, and hence the algorithm makes no attempt to minimize W. We also observe that one can obtain reasonable tradeoffs between area and wire length without substantially increasing the area. Similar observation can be made for the other two pairs of problems. For some of the test problems with nonzero λ, there are solutions in which the value of $A + \lambda W$ is about the same while the value of A is smaller and the value of W is larger. This is to be expected because of the tradeoffs between A and W in the cost function. Also, for problems with zero λ, even though the final wire lengths are large, they are much smaller than those in random solutions. This is because modules that are closely packed will in general lead to a reduction of total wire length too. For some of the test problems, suboptimality in area is due to the presence of very rigid modules (small s_i), and the tradeoffs between area and wire length. Fig.3.9 shows the final floorplan for problem P8 which contains 40 modules.

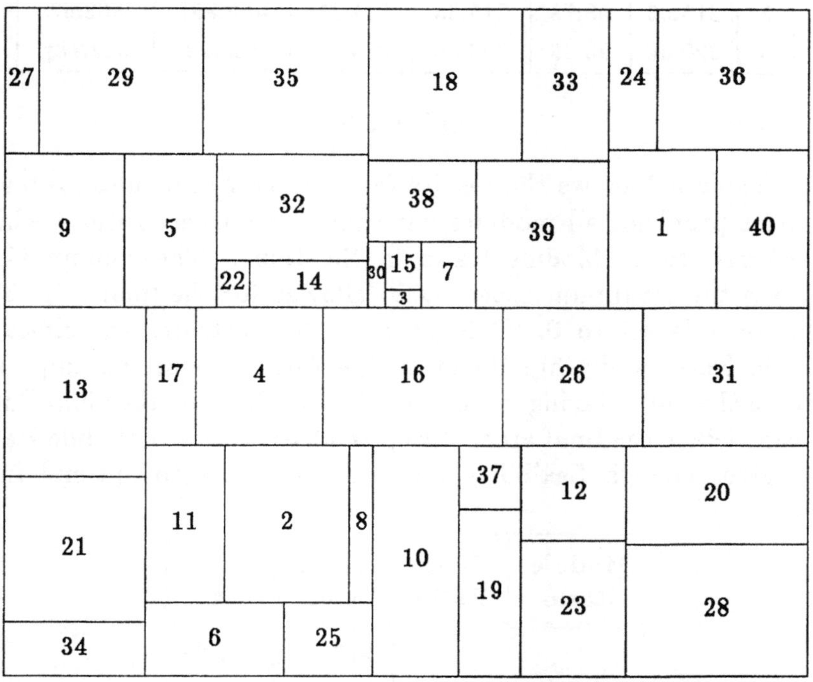

34 21 + 13 + 6 25 * 11 2 * 8 * + 10 * 19 37 + * 23 12 + * 28 20 + * 17 4 * 16 * 26 * 31 * + * 9 5 * 22 14 * 32 + * 27 29 * 35 * + 30 3 15 + * 7 * 38 + 39 * 18 33 * + * 1 40 * 24 36 * + * +

Fig. 3.9 Final floorplan for problem P8.

Table 3.2 is a summary of results for a random problem with 20 modules. The sum of the areas of the modules in this problem is 195.37. Various values of λ are used to demonstrate the trade-offs between area and wire length. As the value of λ increases from 0 to 3, A increases from 196.42 to 220.30 while W decreases from 152.1 to 93.96. Also note that those entries in the table marked by a † are the values of the cost function. They should be smaller than other values in the same column. This is indeed the case.

	S.A. Solution					
λ_w	A	W	A	A + W	A + 2W	A + 3W
0	196.42	152.1	196.42†	348.52	500.62	652.72
1	206.60	103.7	206.60	310.30†	414.00	517.70
2	215.33	95.78	215.33	311.11	406.89†	502.67
3	220.30	93.96	220.30	314.26	408.22	502.18†

Table 3.2

Table 3.3 shows the results for another 20 modules problem. For this problem, all modules have the same aspect ratio s which is referred to as Module Ratio in the table. The column Chip Ratio is the maximum aspect ratio allowed for the final chip. The value of λ is set to 0. This problem demonstrates the effect of Module Ratio and Chip Ratio on the final area of the chip. We observe that by relaxing either the Module Ratio or the Chip Ratio we can reduce the final area. Also, we observe that a Module Ratio of 2 gives enough flexibility for achieving close to optimal final area.

Module Ratio	Chip Ratio	Total Area	A
1	1	229.17	291.23
2	1	–	231.65
2	2	–	230.52
3	3	–	230.19

Table 3.3

3.3. Part 2: Rectangular and L–Shaped Modules

In this section, we present an algorithm to produce floorplans for rectangular and L–shaped modules. (Fig.3.10 shows examples of L–shaped module.) This algorithm is similar in spirit to the algorithm described in the last section. It also uses the search method of simulated annealing, similar floorplan representation (Polish expressions with new operators), and similar ways to locally modify a floorplan. However, not only can the algorithm handle L–shaped modules, in the case where all the modules are rectangular, but it will, in general, also be able to produce non–slicing rectangular floorplans.

For each module, there is a given set of possible shapes, dimension, and orientation for the module. Each possible choice of shape, dimensions, and orientation for a module is called an *instance* of the module. (Clearly, a non–rotatable rigid module has only one instance.) In the last section, we assume that all modules are rectangular. Also, we consider only slicing floorplans there. In that case, the technique of adding up shape curves allows us to simultaneously consider all possible instances of the modules efficiently. The situation in this section is slightly more complicated because there are L–shaped modules and therefore it is necessary to consider more general type of floorplans. The idea of adding up shape curves is no longer applicable. Instead, we shall examine the instances of a module one at a time and shall rely on probabilistic techniques to make selections from the different instances.

Let $\mathbf{\Omega}$ denote the set of rectangular and L–shaped geometric figures of all possible sizes and dimensions. For each module i, let $\mathbf{G}_i$, $\mathbf{G}_i \subseteq \mathbf{\Omega}$, denote the set of all possible instances for the module. We shall define four binary operators and one unary operator that

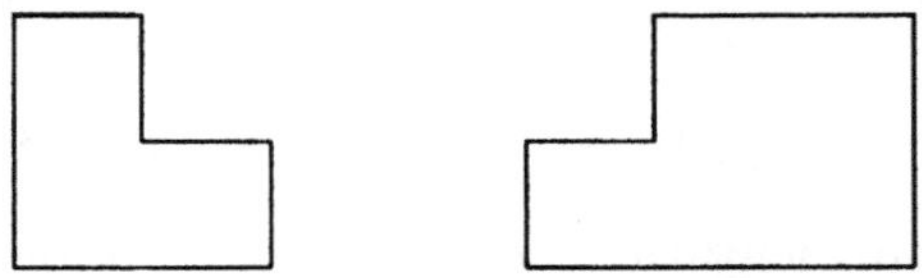

Fig. 3.10 Examples of L–shaped modules.

operate on the geometric figures in Ω. Every algebraic expression in the algebraic system formed by geometric figures chosen from the G_i's corresponds to a floorplan. These algebraic expressions can be represented by Polish expressions. The method of simulated annealing is then used to search for an optimal floorplan among these Polish expressions.

3.3.1. Geometric Figures

We first introduce a way to represent the geometric figures in Ω. Let A be a geometric figure in Ω. The *orientation index* of A is defined to be 0 if A is rectangular, and is either 1, 2, 3, or 4 as shown in Fig.3.11 if A is L-shaped. It it clear that the dimensions of A are completely specified by the length of its four outermost boundary edges. Let x_1 and x_2, $x_1 \geq x_2$, be the lengths of the two outermost horizontal edges. Let y_1 and y_2, $y_1 \geq y_2$, be the lengths of the two outermost vertical edges. Let s be the orientation index of A. Then A can be represented by the 5-tuples (x_1,x_2,y_1,y_2,s). (See Fig.3.12.) Consequently, the set of all rectangular geometric

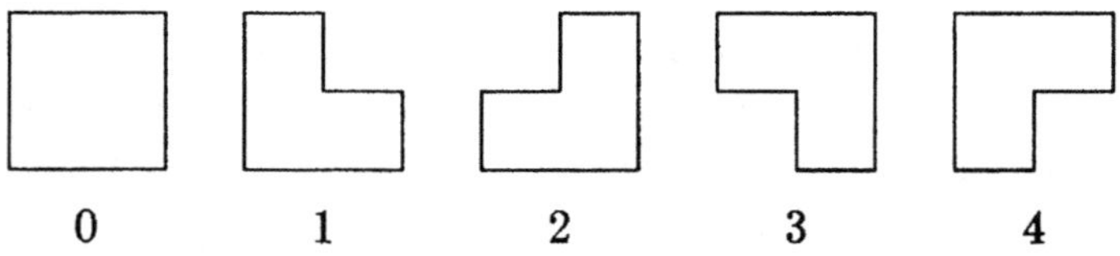

Fig. 3.11 Orientation indices of geometric figures in Ω.

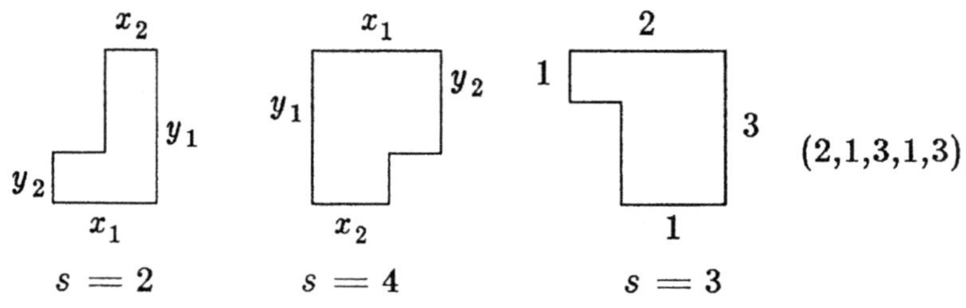

Fig. 3.12 Representation of geometric figures by 5-tuples (x_1, x_2, y_1, y_2, s).

figures can be represented by $\mathbf{\Omega}_1 = \{(x_1,x_2,y_1,y_2,0) \mid x_1 = x_2,\, y_1 = y_2\}$, and the set of all L-shaped geometric figures can be represented by $\mathbf{\Omega}_2 = \{(x_1,x_2,y_1,y_2,s) \mid x_1 \geq x_2,\ y_1 \geq y_2,\ (x_1-x_2)(y_1-y_2) > 0,$ and $s \in \{1,2,3,4\}\}$. The condition $(x_1-x_2)(y_1-y_2) > 0$ excludes the possibility of the geometric figure being rectangular. Hence there is a 1–1 correspondence between $\mathbf{\Omega}$ and the set $\mathbf{\Omega}_1 \cup \mathbf{\Omega}_2$.

3.3.2. The Operators

As was pointed out in Section 3.2.2, slicing floorplans can be obtained by recursively combining rectangles to form larger rectangles by means of two binary operators $*$ and $+$. For two rectangles A and B, $A + B$ and $A * B$ are the rectangles obtained by placing B on top of A, and B to the right of A, respectively. Since we are now dealing with both rectangular and L-shaped geometric figures, a natural extension is to define operators that combine rectangular and L-shaped geometric figures to form larger rectangular and L-shaped geometric figures. We shall define an unary operator $\neg$, and four binary operators $*_1$, $*_2$, $+_1$, and $+_2$ that operate on the geometric figures in $\mathbf{\Omega}$.

Fig.3.13 shows the definition of the unary operator $\neg$ which is a function from $\mathbf{\Omega}$ to $\mathbf{\Omega}$. For $A \in \mathbf{\Omega}$, $\neg A$ is defined to be the smallest bounding rectangle of A.

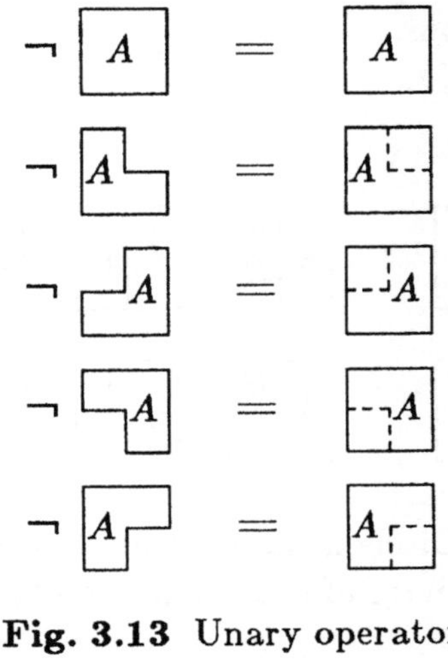

Fig. 3.13 Unary operator.

We shall refer $\neg A$ as the *completion* of A. More precisely, we have $\neg\,(x_1,x_2,y_1,y_2,s) = (x_1,x_1,y_1,y_1,0)$. The unary operator $\neg$ provides the possibility of placing an L–shaped geometric figure in a rectangular region. The four binary operators, which are functions from $\mathbf{\Omega} \times \mathbf{\Omega}$ to $\mathbf{\Omega}$, represent different ways of putting two geometric figures in $\mathbf{\Omega}$ together, as compactly as possible, to form a larger geometric figure in $\mathbf{\Omega}$. Let A, $B \in \mathbf{\Omega}$. Then $A *_1 B$ and $A *_2 B$ are geometric figures obtained by putting A and B next to each other horizontally with B placed to the right of A, and $A +_1 B$ and $A +_2 B$ are geometric figures obtained by putting A and B next to each other vertically with B placed on top of A. The operators are defined in such a way that the rules for combining the geometric figures are completely determined by the orientation indices of the two geometric figures involved. The exact dimensions and orientation index of the new geometric figure are determined by the dimensions of the two geometric figures involved such that packing is as compact as possible. Hence, each binary operator can be completely described by its actions on the 25 possible combinations of the orientation indices of the two operands (5 possible orientation indices for A and 5 possible orientation indices for B). Thus, for the four binary operators, there are all together 100 possibilities. Fig.3.15 shows all of the 100 possible ways of combining two geometric figures based solely on their orientation indices. (In Fig.3.15, each circle at a T–intersection point corresponds to three possible outcomes for the resulting geometric figure. Fig.3.14 illustrates this notation.) We shall give three examples to illustrate these binary operations.

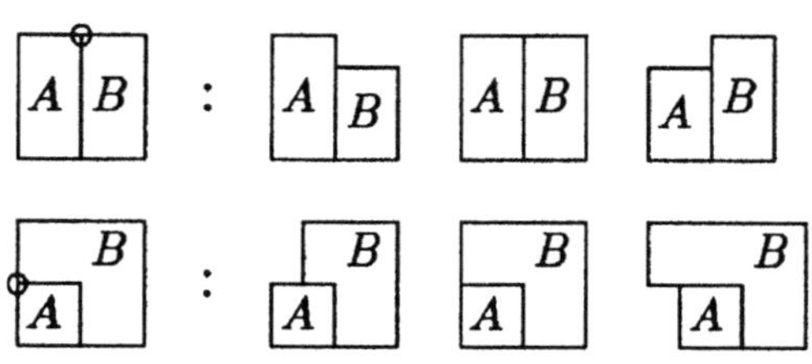

Fig. 3.14 Circle notation representing 3 possible outcomes of combining two figures.

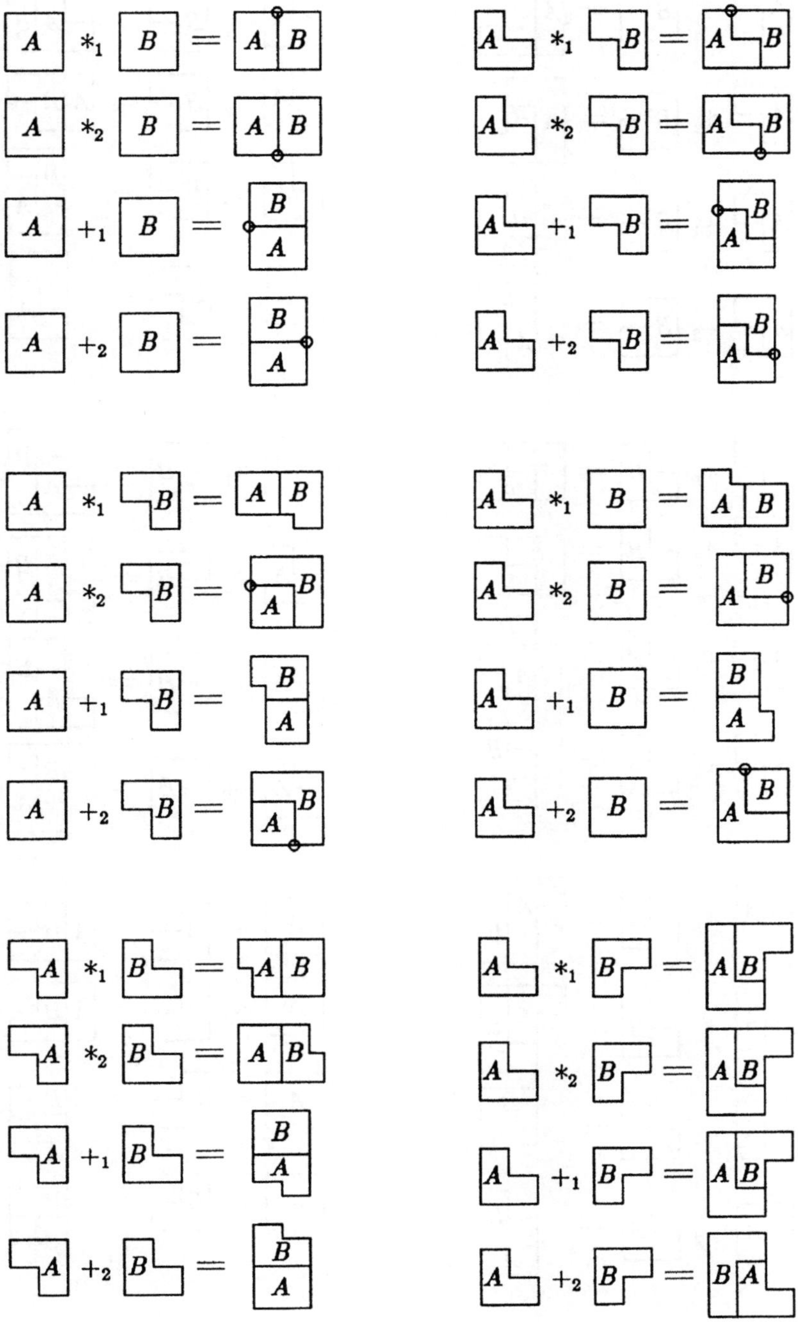

Fig. 3.15 Binary operators.

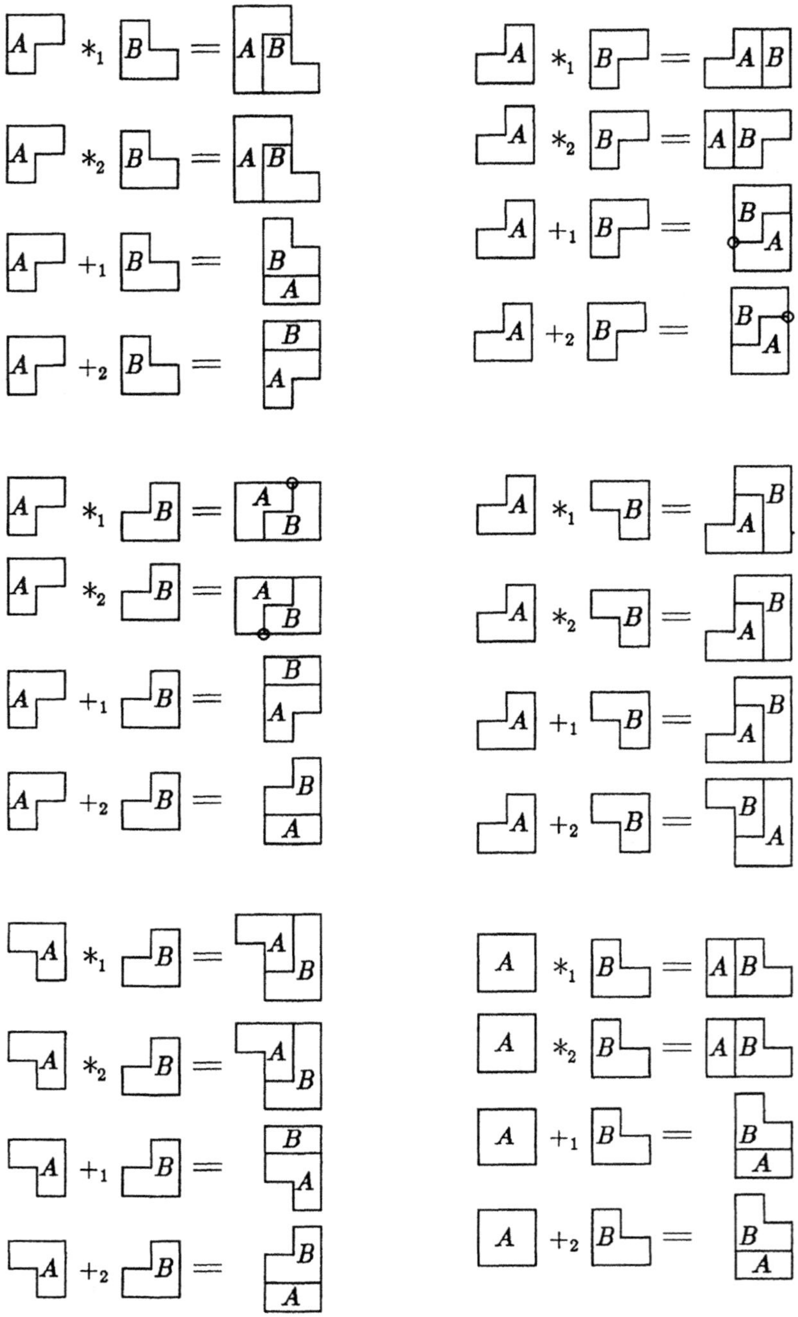

Fig. 3.15 (continued)

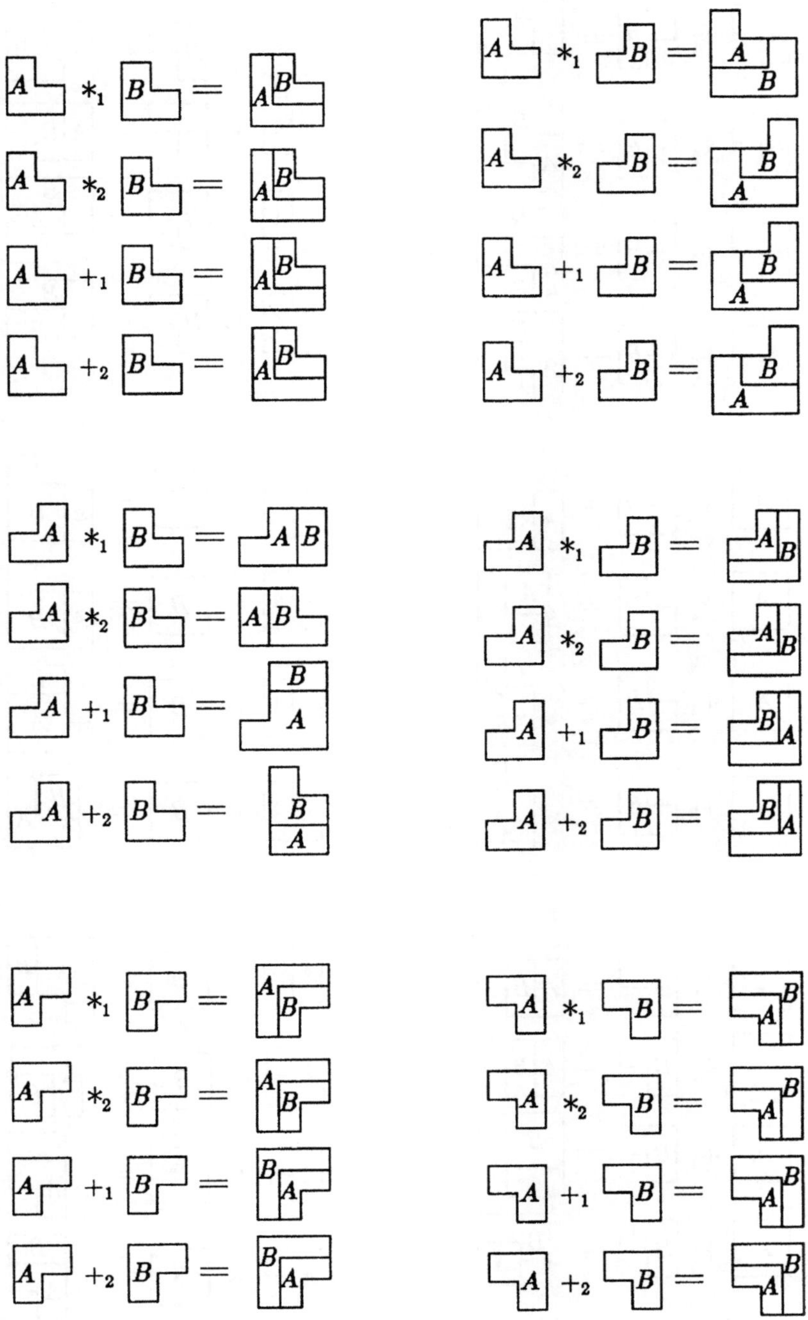

Fig. 3.15 (continued)

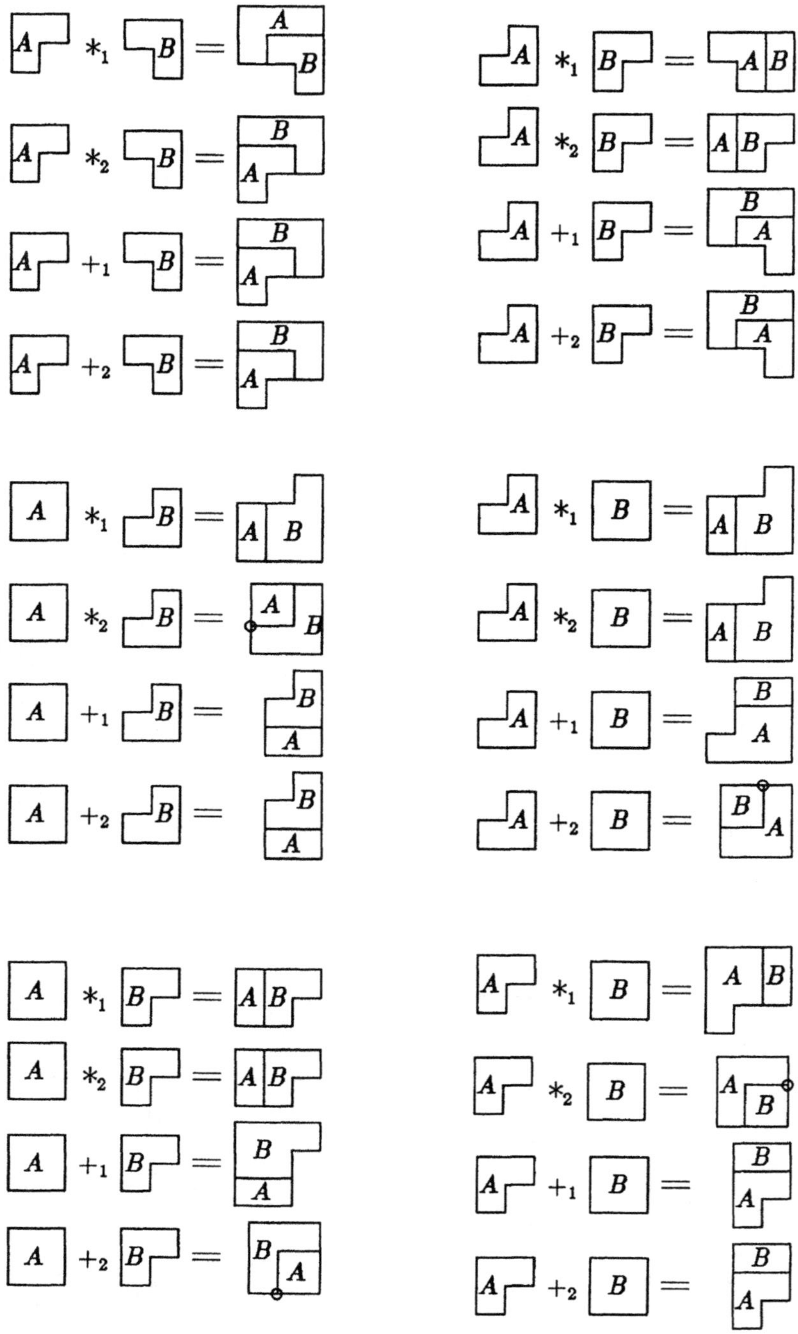

Fig. 3.15 (continued)

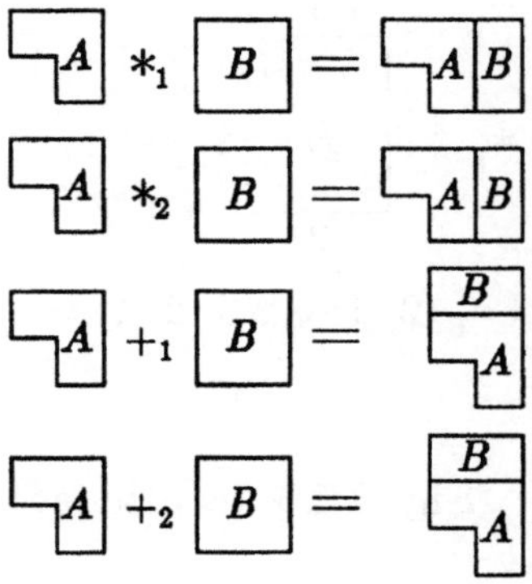

Fig. 3.15 (continued)

Example 1. We consider here the case of $A *_1 B$ when the orientation indices of A and B are 3 and 1, respectively. (See Fig.3.16(a).) Let $A = (a_1,a_2,b_1,b_2,3)$ and $B = (c_1,c_2,d_1,d_2,1)$. We have $A *_1 B = (x_1,x_2,y_1,y_2,s)$ where the values of x_1, x_2, y_1, y_2, and s are determined by the following procedure:

$$
\begin{aligned}
x_1 &\leftarrow a_1 + c_1; \\
x_2 &\leftarrow a_2 + c_1; \\
y_1 &\leftarrow \max\,(b_1,d_1); \\
y_2 &\leftarrow b_2; \\
s &\leftarrow 3;
\end{aligned}
$$

For example, we have $(2,1,2,1,3)\ *_1\ (2,1.5,2.5,2,1) = (4,3,2.5,1,3)$. (See Fig.3.16(b) for a pictorial illustration of this computation.)

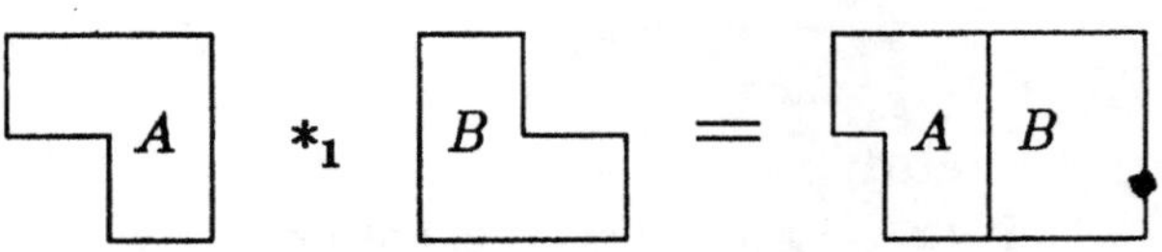

(a)

Fig. 3.16 An example of a binary operation.

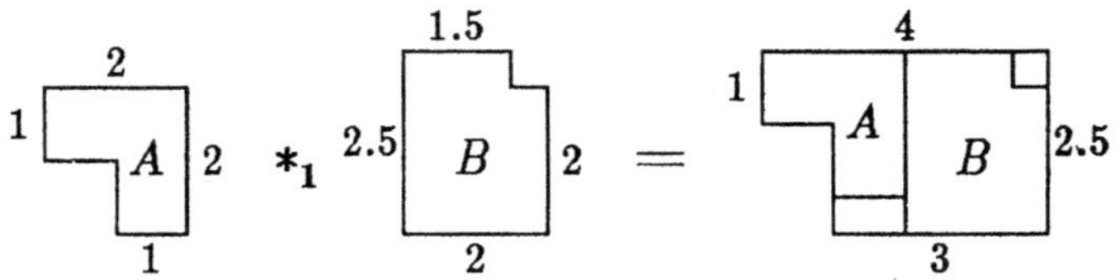

$(2, 1, 2, 1, 3) *_1 (2, 1.5, 2.5, 2, 1) = (4, 3, 2.5, 1, 3)$

(b)

Fig. 3.16 (continued)

Example 2. We consider here the case of $A +_2 B$ when the orientation indices of A and B are 1 and 0, respectively. (See Fig.3.17(a).) Let $A = (a_1,a_2,b_1,b_2,1)$ and $B = (c_1,c_1,d_1,d_1,0)$. We have $A +_2 B = (x_1,x_2,y_1,y_2,s)$ where the values of x_1, x_2, y_1, y_2, and s are determined by the following procedure:

```
x1 ← max (a1,a2+c1);
y1 ← max (b1,b2+d1);
if (y1 > b1) then
  begin
    x2 ← c1;
    y2 ← b1;
    s  ← 2;
  end
else
  begin
    x2 ← a2;
    y2 ← b2 + d1;
    s  ← 1;
  end;
if (x1 = x2) or (y1 = y2) then
  (x1,x2,y1,y2,s) ← (x1,x1,y1,y1,0);
```

The last **if–then** statement is for the case in which the resulting geometric figure is a rectangle because of the dimensions of A and B. (See Fig.3.17(b) for a pictorial illustration of the computation $(6,2,4,2,1) +_2 (3,3,3,3,0) = (6,3,5,4,2)$.)

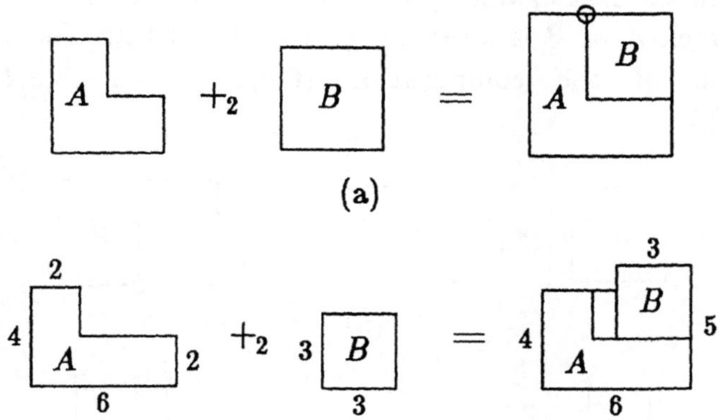

$(6, 2, 4, 2, 1) +_2 (3, 3, 3, 3, 0) = (6, 3, 5, 4, 2)$

(b)

Fig. 3.17 An example of a binary operation.

Example 3. We consider here the case of $A *_2 B$ when the orientation indices of A and B are both 0. (See Fig.3.18(a).) Let $A = (a_1, a_1, b_1, b_1, 0)$ and $B = (c_1, c_1, d_1, d_1, 0)$. We have $A *_2 B = (x_1, x_2, y_1, y_2, s)$ where the values of x_1, x_2, y_1, y_2, and s are determined by the following procedure:

```
x1 ← a1 + c1;
y1 ← max (b1,d1);
if (y1 > d1) then
  begin
    x2 ← a1;
    y2 ← d1;
    s  ← 4;
  end
else
  begin
    x2 ← c1;
    y2 ← b1;
    s  ← 3;
  end;
if (x1 = x2) or (y1 = y2) then
  (x1,x2,y1,y2,s) ← (x1,x1,y1,y1,0);
```

As in the last example, the last **if–then** statement is for the case in which $A *_2 B$ is a rectangle. (See Fig.3.18(b) for a pictorial illustration of the computation $(6,6,2,2,0) *_2 (3,3,4,4,0) = (5,2,6,4,4)$.)

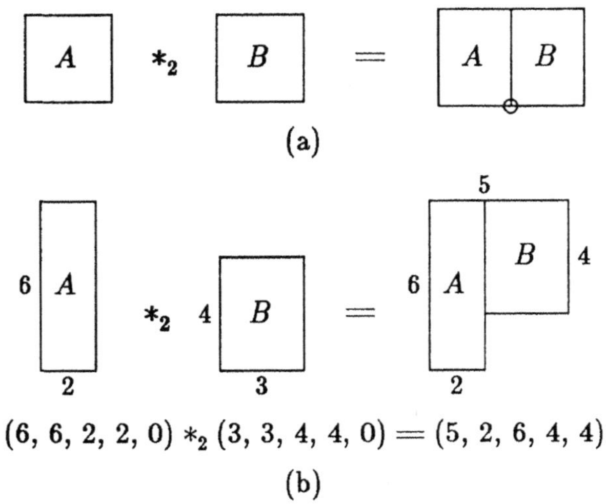

Fig. 3.18 An example of a binary operation.

It should also be noted that the operators constitute a *complete* set in the sense that we can generate all minimally compacted combinations of geometric figures. More precisely, we have the following theorem.

Theorem 3.5. Let $f : \mathbf{\Omega} \times \mathbf{\Omega} \rightarrow \mathbf{\Omega}$ be any function that combine two geometric figures to yield a larger geometric figure. For any $A, B \in \mathbf{\Omega}$, there exists $C \in \mathbf{\Omega}$ such that C is small enough to be placed inside $f(A,B)$, and C is of the form $A' \circ B'$ or $B' \circ A'$, where A' is either A or $\neg A$, B' is either B or $\neg B$, and $\circ \in \{*_1, *_2, +_1, +_2\}$.

Proof. We assume that the orientation index of $f(A,B)$ is 1. The proof for the other cases are similar. $f(A,B)$ is an L–shaped region containing two non–overlapping geometric figures A and B. The first step of the proof is to complete both A and B as much as possible. If $\neg A$ remains inside $f(A,B)$ and does not intersect B, we let A' be $\neg A$, otherwise we let A' be A. Similarly, if $\neg B$ remains inside $f(A,B)$ and does not intersect A', we let B' be $\neg B$, otherwise we let B' be B. There are several cases to be considered.

Case 1: Both A' and B' are rectangles. Suppose there is a vertical line separating A' and B'. Without loss of generality, we may assume that A' is to the left of B'. C is the first geometric figure shown in Fig.3.19(a). If no such vertical line exists, there must be a horizontal line separating A' and B'. Again, we may assume that A' is below B'. C is the second geometric figure shown in Fig.3.19(a). (Note that because of the orientation index of $f(A,B)$ is 1, we need not consider the other two combining rules where the circle symbol appears either on the left side or at the bottom. This also applies to the other cases.)

Case 2: One of A' and B' is a rectangle, and the other is L–shaped. We may assume that A' is L–shaped. If $\neg A'$ intersects B', C can be chosen from the figures shown in Fig.3.19(b). Otherwise, we have $\neg A'$ intersects the outside of $f(A,B)$ and C can be chosen from the figures shown in Fig.3.19(c).

Case 3: Both A' and B' are L–shaped. Suppose $\neg A'$ intersects B' and $\neg B'$ intersects A'. We may assume that A' is either to the left of B' or below B'. C can be chosen from those shown in Fig.3.19(d). If the completions of A' and B' are not mutual intersecting, we may assume that $\neg A'$ intersects B', $\neg B'$ intersects the outside of $f(A,B)$ and $\neg B'$ does not intersect A'. In this case, C can be chosen from the figures shown in Fig.3.19(e).

Note that the southwest corner of C is well defined except when C is given by the last geometric figure in Fig.3.19(b). In that case, we use the southwest corner of $\neg C$ instead. Now if we place the southwest corner of C at the southwest corner of $f(A,B)$, C stays completely inside $f(A,B)$. The theorem follows from the fact that all the geometric figures in Fig.3.19 can be generated by the operators. □

Fig.3.20 shows three examples of $f(A,B)$ together with the corresponding smaller geometric figures generated by the operators.

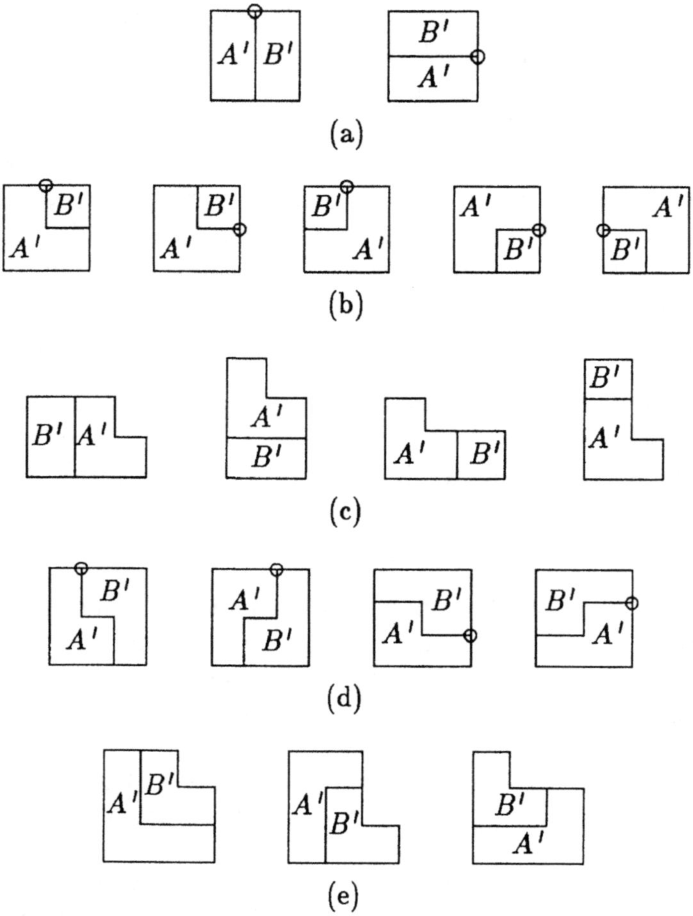

Fig. 3.19 Possible choices for C (proof of Theorem 3.5).

3.3.3. Floorplan Representation

Consider the *algebraic system* $(\mathbf{\Omega}, *_1, *_2, +_1, +_2, \neg)$. It is clear that an *algebraic expression* in $(\mathbf{\Omega}, *_1, *_2, +_1, +_2, \neg)$ represents a way to combine a set of geometric figures in $\mathbf{\Omega}$ to yield a resultant geometric figure in $\mathbf{\Omega}$. Consider as an example the algebraic expression $(\neg A +_1 B) *_1 (\neg(C *_2 D) +_2 E)$ with A, B, C, D, $E \in \mathbf{\Omega}$. (The order of precedence of the binary operations is specified by parentheses and we use the convention that unary operation has priority over binary operations.) Fig.3.21 shows the geometric figure corresponding to the expression.

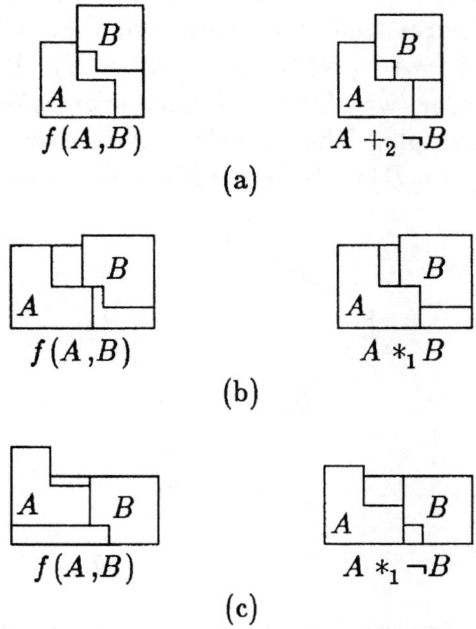

Fig. 3.20 Compact geometric figures generated by the operators.

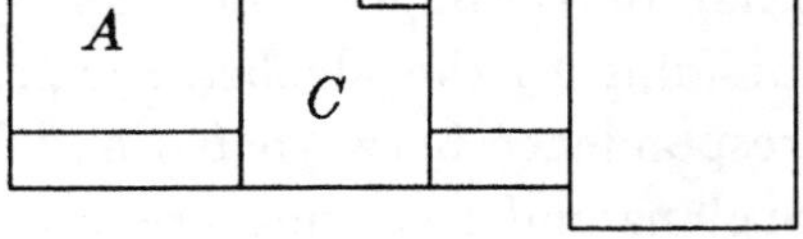

Fig. 3.21 Geometric figure corresponding to $(\neg A +_1 B) *_1 (\neg (C *_2 D) +_2 E)$.

Similar to the case discussed in Section 3.2, an algebraic expression γ can be represented by a tree $\mathbf{T}$. The internal nodes of $\mathbf{T}$ correspond to the operators in γ and the leaf nodes correspond to the geometric figures in γ. Let v be an internal node of $\mathbf{T}$. If v is a binary operator, then v has two subtrees. These subtrees correspond to the two geometric figures that v combines. If v is the unary operator $\neg$, then v has one subtree which corresponds to the geometric figures that u operates on. Fig.3.22 shows the tree representation of $(\neg A +_1 B) *_1 (\neg(C *_2 D) +_2 E)$. If we traverse the tree $\mathbf{T}$ in postorder, we obtain a *Polish expression* representation of γ. For example, the Polish expression obtained from $(\neg A +_1 B) *_1 (\neg(C *_2 D) +_2 E)$ is $A \neg B +_1 CD *_2 \neg E +_2 *_1$.

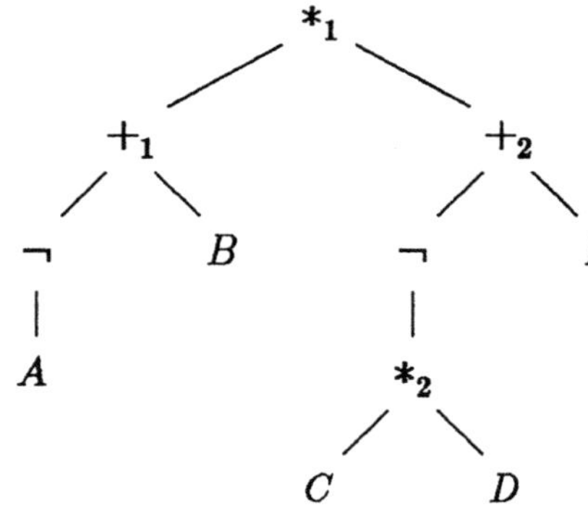

Fig. 3.22 Tree representation for $(\neg A +_1 B) *_1 (\neg(C *_2 D) +_2 E)$.

Let $\mathbf{E}$ be the set of all algebraic expressions with *operands*[1] $m_1, m_2, \ldots, m_n$, where $m_i \in \mathbf{G}_i$, $1 \leq i \leq n$. The set $\mathbf{E}$ represents different ways of compactly combining the given set of modules. Clearly, each algebraic expression $\boldsymbol{\alpha}$ in $\mathbf{E}$ represents a floorplan for the set of given modules. (More precisely, $\neg \boldsymbol{\alpha}$, the completion of $\boldsymbol{\alpha}$, represents a floorplan.) Furthermore, the regions in the floorplan are either rectangular or L–shaped. Let $\mathbf{E_P}$ be the set of Polish expressions corresponding to the algebraic expressions in $\mathbf{E}$. Since there is a 1–1 correspondence between $\mathbf{E_P}$ and $\mathbf{E}$, we may assume that the set of floorplans under consideration are given by $\mathbf{E_P}$. We shall search for an optimal floorplan among all the Polish expressions in $\mathbf{E_P}$.

---

[1]We shall refer the elements in $\boldsymbol{\Omega}$ that appear in an algebraic expression to be the *operands* of the expression.

3.3.4. The Algorithm

Let $\boldsymbol{\alpha}$ be a Polish expression and $\mathbf{f}_{\boldsymbol{\alpha}}$ be the floorplan represented by $\boldsymbol{\alpha}$. Let $A(\boldsymbol{\alpha})$ be the area of $\mathbf{f}_{\boldsymbol{\alpha}}$. In other words, if $(x_1,x_2,y_1,y_2,s) \in \boldsymbol{\Omega}$ is the geometric figure corresponding to $\boldsymbol{\alpha}$, $A(\boldsymbol{\alpha}) = x_1y_1$. (In the case when there are two numbers p and q specifying the allowed range of aspect ratio of the final chip, $A(\boldsymbol{\alpha})$ is defined as the area of the smallest rectangle, with aspect ratio between p and q, containing (x_1,x_2,y_1,y_2,s).) Let $W(\boldsymbol{\alpha})$ be the total wire length of $\mathbf{f}_{\boldsymbol{\alpha}}$. The cost function is $\Psi(\boldsymbol{\alpha}) = A(\boldsymbol{\alpha}) + \lambda W(\boldsymbol{\alpha})$, where λ is a given constant that controls the relative importance of area and wire length.

The floorplan design algorithm employs the method of simulated annealing to search for a Polish expression $\boldsymbol{\alpha}$ in $\mathbf{E_P}$ such that $\Psi(\boldsymbol{\alpha})$ is minimized. Let $\boldsymbol{\alpha} \in \mathbf{E_P}$ be a Polish expression. It is clear that $\boldsymbol{\alpha}$ is of the form $\alpha_1\beta_1\alpha_2\beta_2 \cdots \alpha_{2n-1}\beta_{2n-1}$ where n of the α_i's are geometric figures for the n modules (operands), the other n α_i's are binary operators, and each β is either the unary operator $\neg$ or the empty string ϵ. There are four types of moves that can be used to locally modify $\boldsymbol{\alpha}$. Two Polish expressions $\boldsymbol{\alpha}$ and $\boldsymbol{\alpha}'$ in $\mathbf{E_P}$ are said to be *neighbors* if one can be obtained from the other via one of these moves. The four types of moves are defined as follows:

M1. Modify α_i for some i.
M2. Modify β_i for some i.
M3. Swap two operands.
M4. Swap two adjacent operand and binary operator in $\alpha_1\alpha_2 \cdots \alpha_{2n-1}$.

We now give a more detailed description of the moves. For M1 moves, there are two cases corresponding to whether α_i is an operand or a binary operator. If α_i is an operand, we have $\alpha_i = A \in \mathbf{G}_k$ for some k. We set $\alpha_i \leftarrow A'$ where $A' \in \mathbf{G}_k$ and $A \neq A'$. This corresponds to selecting another instance for a module. (Note that if $|\mathbf{G}_k| = 1$, no modification is possible.) On the other hand, if α_i is a binary operator, we select a different binary operator for α_i. For M2 moves, we change β_i to the other element in $\{\epsilon,\neg\}$. Note that in a Polish expression $\boldsymbol{\alpha}$, each operand represents a geometric figure of a module, and the subtree rooted at each operator represents a geometric figure for a "super-module." We shall refer to these geometric figures as the geometric figures represented by the α_i's and the β_i's in $\boldsymbol{\alpha}$. Clearly, M2 has no effect at all when the geometric figure represented by α_i is a rectangle. Therefore, we only modify those β_i where the geometric figure represented by α_i is L-shaped. For M3 moves, we swap two operands α_i and α_j. This corresponds to swapping two modules. Finally, M4 moves correspond to swapping α_i and α_{i+1} for some i when one of them is an operand and the other is a binary operator. Note that M1, M2, or M3 moves always result in another Polish expression in $\mathbf{E_P}$, but M4 moves might produces an expression which is not a Polish expression. For M4 moves, only those that produce a valid Polish expression are considered. As in Section 3.2, the four types of moves are sufficient to ensure that it is possible to transform any Polish expression into any other via a sequence of moves. Again, there is a sequence of $O(n^2)$ neighboring Polish expressions between any two Polish expressions.

The temperature schedule is of the form $T_k = r \cdot T_{k-1}$, $r = 1, 2, 3, \ldots$, where r is usually between 0.8 and 0.9. In the current implementation, the stopping criteria are: (1) the number of accepted moves in that temperature is very small; or (2) the average cost remains unchanged for three consecutive temperatures.

We should point out that throughout the entire annealing process, we need to compute the cost of an algebraic expression many times. Since each move is only a minor modification of a Polish expression, recomputations should be avoided as much as possible. Recall that in a Polish expression $\boldsymbol{\alpha}$, each α_i and each β_i represents a geometric figure for either a module or a "super-module." Computation of the cost of $\boldsymbol{\alpha}$ amounts to determining all these geometric figures. Fortunately, after every move, many of these geometric figures remain unchanged. The following theorem characterizes the set of geometric figures that needs to be recomputed. The proof of the theorem is very similar to that of Theorem 3.4 in Section 3.2.

Theorem 3.6. Let $\boldsymbol{\alpha}'$ be the expression obtained from $\boldsymbol{\alpha}$ after a move. Let $\mathbf{T}_{\boldsymbol{\alpha}}$ and $\mathbf{T}_{\boldsymbol{\alpha}'}$ be the trees corresponding to $\boldsymbol{\alpha}$ and $\boldsymbol{\alpha}'$, respectively. The geometric figures that need to be recomputed correspond to subtrees rooted at operators that lie on one or two paths in each of $\mathbf{T}_{\boldsymbol{\alpha}}$ and $\mathbf{T}_{\boldsymbol{\alpha}'}$.

Finally, we note that for the case in which the given modules are rectangular, the algorithm is capable of producing a non-slicing floorplan. As an example, we note that the algebraic expression $((B +_2 A) +_2 C) *_1 (D +_1 E)$ corresponds to the non-slicing floorplan shown in Fig.3.23.

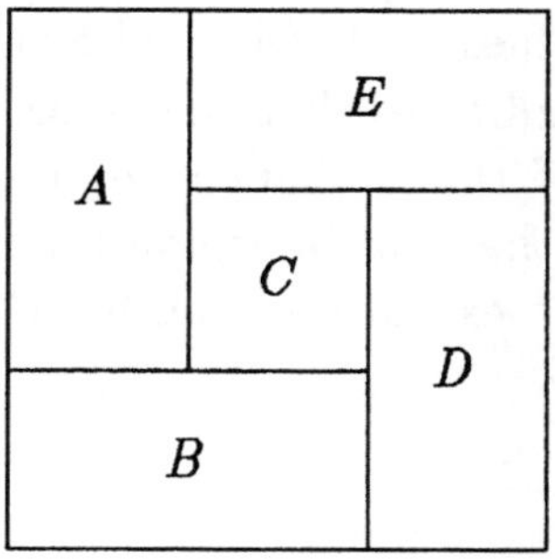

Fig. 3.23 A non-slicing floorplan.

3.3.5. Experimental Results

The algorithm has been implemented in Pascal on a PYRAMID computer. The performance of the algorithm is compared with that of the algorithm in Part 1. Since the algorithm in Part 1 is designed to handle only rectangular modules and only produces slicing floorplans, hence, in order to use that algorithm, each L-shaped module is replaced by its smallest bounding rectangle when the algorithm is tested. The results are summarized in Table 3.4 where n is the number of modules and δ measures the

Problem	n	δ	A_L/A_S
Q1	20	0.81	0.85
Q2	20	0.84	0.89
Q3	25	0.85	0.88
Q4	40	0.83	0.86
Q5	25	0.66	0.75

Table 3.4

amount of dead space introduced by replacing all L–shaped modules by their bounding rectangles. More precisely, if A is the total area of all the modules and A' is the total area of the bounding rectangles of all the modules, $\delta = A/A'$. A_S is the area of the floorplan produced by the algorithm in Part 1 and A_L is the area of the floorplan produced by the algorithm described in this section. Note that the values of A_S/A_L and δ are quite close for all the test problems. This indicates that the algorithm in this section can take full advantage of the L–shaped modules to reduce total chip area. Fig.3.24 shows the final floorplan for problem Q4 which contains 40 modules. The execution time for this problem was about 10 CPU minutes.

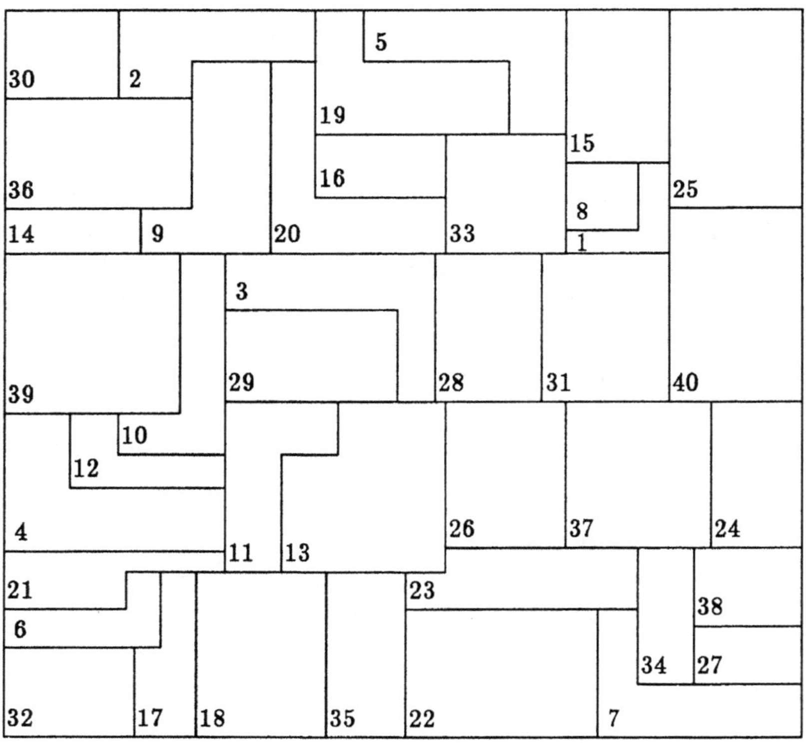

Fig. 3.24 Final floorplan for problem Q4.

3.4. Concluding Remarks

We presented in this chapter two algorithms for the floorplan design problem. The algorithms are similar in spirit. They both use Polish expressions to represent floorplans, employ the search method of simulated annealing, and use similar ways to locally modify a floorplan. The first algorithm is for the case where all modules are rectangular. All floorplans produced by this algorithm are slicing floorplans. The second algorithm is for the case where the modules are either rectangular or L–shaped. This algorithm is capable of producing a non–slicing floorplan when all the modules are rectangular. Both algorithms consider simultaneously the interconnection information as well as the area and shape information. Experimental results indicate that the algorithms perform well for many test problems.

CHAPTER 4

CHANNEL ROUTING

4.1. Introduction

The notion of *channel routing* was first introduced by Hashimoto and Stevens in 1971 [HaSt71]. Since then it has been used extensively in the layout of integrated circuits. This is an effective approach to the general routing problem since it leads to a decomposition of the overall routing problem into a number of smaller sub-problems that can be solved independently. Furthermore, it is flexible enough so that it can be used in various design styles such as gate arrays, standard cells, and macro/custom cells.

In the channel routing problem, we are given two rows of pins across a rectangular channel, and a collection of nets which are sets of pins to be interconnected. Two interconnection layers are available - one for horizontal wire segments and the other for vertical wire segments. Connections between the two layers are made through *via* holes. The objective of the channel router is to interconnect all the nets using a minimum number of tracks. Other parameters such as the number of vias and the lengths of the nets are also used as secondary measures of the quality of a routing solution.

It is well-known that the problem of determining a solution with a minimum number of tracks is NP-complete [Szym85, GaJo79]. However, because of its importance, the channel routing problem has been studied extensively over the past decade and a half, and many efficient channel routers have been developed [HaSt71, KeSP73, Deut76, YoKu82, RiFi82, Chan83, BuPe83, ReSS85]. Although these routers do not guarantee optimal (namely, minimum number of tracks) solutions, the routing solutions they produce are, in practice, either optimal or very close to optimal.

In this chapter, we describe two channel routers [LeWL85] based on the method of simulated annealing. The two channel routers will be referred to as SACR and SACR2. SACR allows either no doglegging or restricted doglegging. SACR2 is an extension of SACR that allows unrestricted doglegging. These routers represent a first example on the application of the method of simulated annealing to *detailed routing*. They compare favorably with the channel router of Yoshimura and Kuh [YoKu82] and that of Rivest and Fiduccia [RiFi82]. In many benchmark problems, they produce optimal results. For the famous "Deutsch's difficult problem," SACR came up with the currently best known result, namely, 28 tracks without doglegs, and 20 tracks with restricted doglegs. SACR2 also came up with a 20 track solution using unrestricted doglegging without split nets.

This chapter is organized as follows. Section 4.2 presents a precise formulation of the problem. SACR is described in detail in Section 4.3, while Section 4.4 describes the extension to SACR2. Experimental results are described in Section 4.5. Finally, concluding remarks are presented in Section 4.6.

4.2. The Channel Routing Problem

We now give a formal description of the channel routing problem. We are given two rows of pins, $T = \{t_1, t_2, \ldots, t_l\}$ and $B = \{b_1, b_2, \ldots, b_l\}$, along the top and bottom sides of a rectangular channel. Each pin t_i or b_i is assigned a *net-number* which is an integer from the set $\{0,1,2,\ldots,n\}$. The set of pins with the same net-number constitutes a *net*, and these pins are to be interconnected with wires. A pin with net-number 0 means that it does not belong to any net (unused). In addition, we are given two sets of nets L and R which are the nets that enter the channel from the left and the right side of the channel, respectively.

We first introduce some terminologies. The connection path for a net consists of *horizontal* and *vertical wire segments*. We assume the *reserved layer model* in which vertical segments lie on one layer and horizontal segments on the other. Horizontal and vertical segments belonging to the same net are connected through *via holes*. However, two vertical (horizontal) wire segments belonging to different nets cannot overlap. These constraints are referred to as *vertical* (*horizontal*) *constraints.* We say that net i *is placed above* net j if the horizontal wire segment of net i is above that of net j. The number of (horizontal) tracks used for routing is called the *width* of the channel.

The objective of the channel router is to interconnect all the nets so that (1) there is no constraint violation, and (2) the width is minimized. Minimizing the width implies minimizing the area, since the length of the channel is usually fixed. In general, other parameters are also important. For example, the total wire length and the number of vias used are some secondary measures used to measure the quality of a channel routing solution. Fig.4.1(a) shows an instance of the channel routing problem while a solution with width 5 is shown in Fig.4.1(b).

Vertical Constraint Graph

Consider the two pins at column i. Suppose pins t_i and b_i belong to different (non-zero) nets, k_1 and k_2, respectively. Then net k_1 *must* be placed above net k_2 so that the vertical wire segments for the two nets will not overlap. In general, the constraints imposed on the nets at the columns are described by a directed graph G, called the *vertical constraint graph*, where each vertex corresponds to a net and a directed edge from net i to net j means that net i must be placed above net j. Fig.4.1(c) shows the vertical constraint graph for the problem shown in Fig.4.1(a).

Density

Define the *span* of a net to be the interval $[l_k, r_k]$ where l_k and r_k are the positions of the leftmost and rightmost pin of net k, respectively. Define d_i, the *local density at column* i, to be the number of nets whose spans intersects column i. (A net with span $[i,i]$ is disregarded in this definition.) The *channel density* of a problem, denoted by d, is the maximum local density over all the columns in the channel.

Lower Bounds

There are two well-known lower bounds on the channel width w; one due to horizontal constraints, and the other due to vertical constraints. The first is the channel density d, namely, $w \geq d$. Such a lower bound is easy to see. Consider a column with

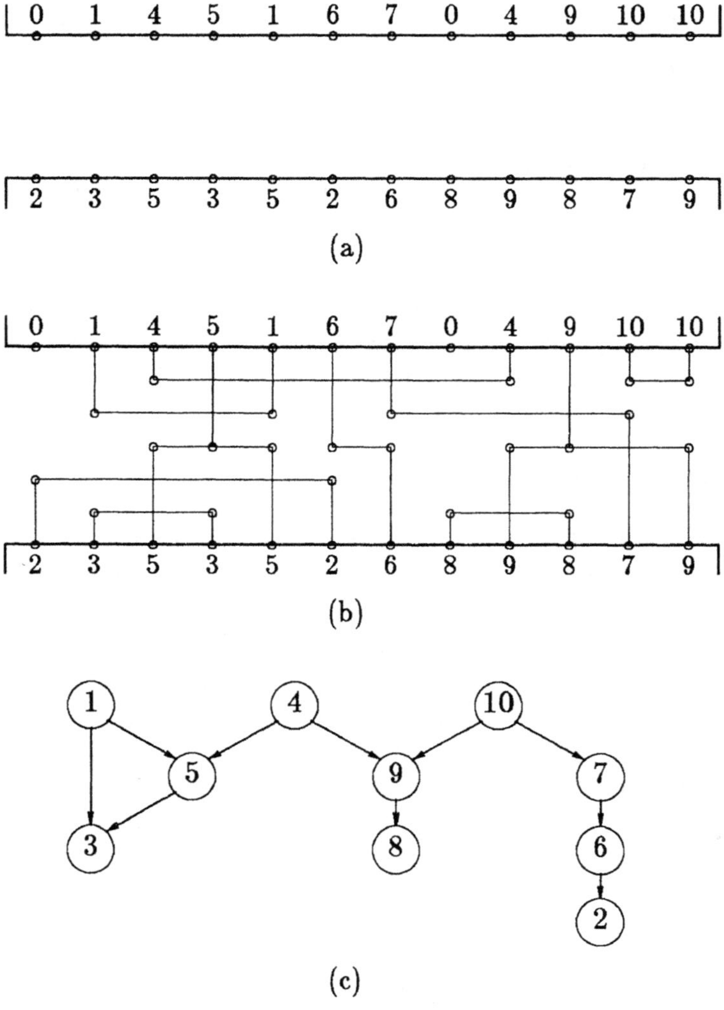

Fig. 4.1 Channel routing. (a) A channel routing problem. (b) A solution to the channel routing problem. (c) The vertical constraint graph.

maximum local density, *i.e* a column with d nets intersecting it. Clearly, no two of these nets can occupy the same horizontal track, and so at least d wiring tracks are needed. Define a *constraint chain* to be a directed path in the vertical constraint graph G. For any constraint chain $(v_1, v_2, \ldots, v_m)$ in G, we must have net v_1 placed above net v_2, which, in turn, must be placed above net v_3, and so on. Thus, the nets $v_1, v_2, \ldots, v_m$ must lie on m distinct tracks. Hence, another lower bound to the width of any solution is the length of the longest constraint chain in G, namely, $w \geq p$, where p is the length of the longest chain in G.

Doglegging

Earlier algorithms for channel routing [HaSt71, KeSP73] assume that a net can only occupy one horizontal track, namely, a net can contain only one horizontal wire segment. Using this assumption, it is easy to see that if there is a cycle in G then no routing solution can be realized. For example, in Fig.4.2(a), net 1 must lie above net 2 because of the vertical constraint at column 1. On the other hand, net 2 must lie above net 1 due to the vertical constraint at column 3. This constitutes a cycle in the vertical constraint graph G. (See Fig.4.2(b).) Clearly, a routing solution cannot be realized unless we allow some of the nets to use more than one horizontal segment as shown in Fig.4.2(c). Such splitting of a net into more than one horizontal segments is called *dogleg-ging.* The notion of doglegging was first introduced by Deutsch [Deut76] in his Dogleg Channel Router. Doglegging not only serves to break constraint cycles but, more importantly, it can also be used to reduce the number of horizontal tracks needed for routing by breaking up long constraint chains. Consider the example in Fig.4.3(a) where an optimal solution requiring four tracks is shown

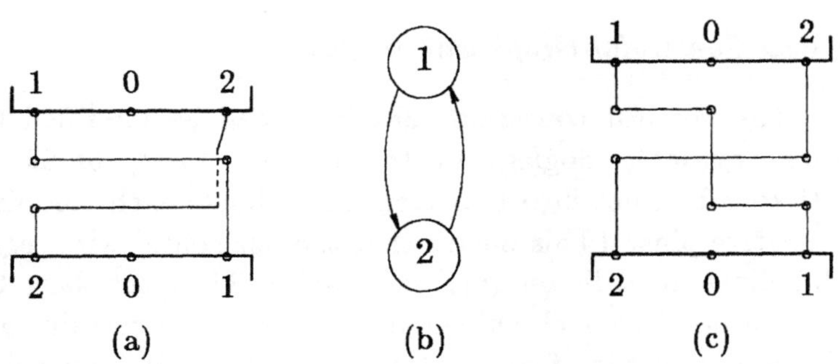

Fig. 4.2 An example with cyclic constraint. (a) No solution. (b) Constraint graph. (c) Solution with dogleg.

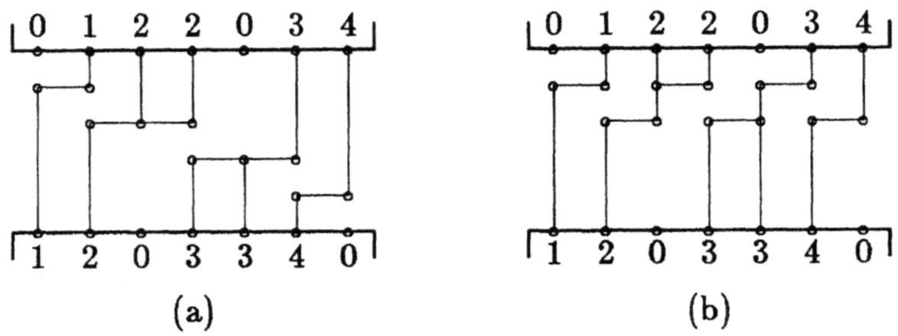

Fig. 4.3 An example illustrating the advantage of using doglegs. (a) Without doglegs. (b) With doglegs.

if doglegging is not allowed. However, with doglegging, the problem can be solved using only two tracks as shown in Fig.4.3(b).

There are two types of doglegging: restricted and unrestricted doglegging. In ***restricted doglegging***, horizontal splitting of a net is allowed only at columns where there is a pin belonging to that net. This is illustrated in Fig.4.3(b) where dogleg for net 2 is allowed only at column 3 and dogleg for net 3 is allowed only at column 5. In the more general case of ***unrestricted doglegging***, a net may be split at any column in the channel. For example, in Fig.4.2(c), a dogleg for net 1 is allowed at column 2 even when there is no pin belonging to net 1 at that column.

Vertical Constraint Graph with Doglegging

The vertical constraint graph G can be modified to incorporate restricted doglegs in the model. First, break up each multi–terminal net into two–terminal subnets each containing two consecutive pins. (This means that the horizontal wire segment for each subnet must lie on only one track.) Then, let there be a vertex correspond to each subnet in the vertical constraint graph G, and there is an edge from vertex i to vertex j if subnet i must lie above subnet j. Fig.4.4(a) gives an example where each subnet is shown as a horizontal line segment corresponding to its span. If we consider the pins on column 5, then subnet 5a must be placed above subnets 4a and 4b. These are indicated by the directed edges (5a,4a) and (5a,4b) in the vertical constraint graph shown in Fig.4.4(b).

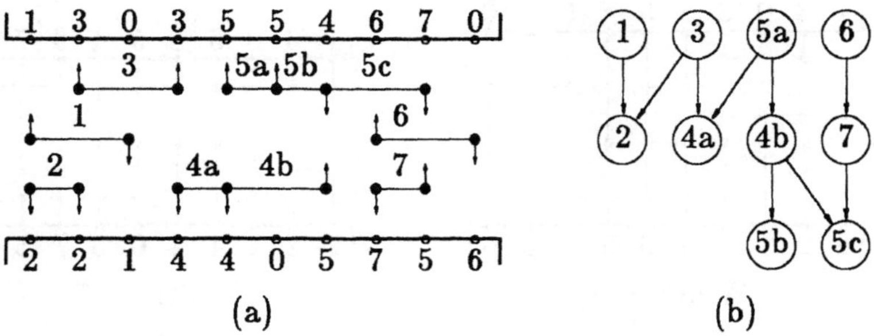

Fig. 4.4 Vertical constraint graph with restricted doglegs.

4.3. The Channel Router SACR

In this section, we describe the simulated annealing channel router SACR. This router only allows unrestricted doglegging. Before describing the details, we first present an overview. The channel router SACR starts with an initial solution. It then repeatedly moves (by minor perturbation of the placement of subnets) from one solution to a neighboring solution. Consider the problem shown in Fig.4.4(a). Assume that at some point, the current solution obtained by SACR is given in Fig.4.5(a). Then Fig.4.5(b)–(d) illustrates the sequence of routing solutions resulting from a sequence of moves made by SACR.

Thus, a major difference between SACR (and SACR2) and all the other channel routers in the literature is that in the other routers, routing solutions are built up incrementally, whereas SACR (and SACR2) works with *complete solutions*, moving from one complete solution to another. Because of this, one special feature of the algorithm is that one can trade–off between the quality of the solution and the running time of the algorithm. If running time is a crucial consideration, one can terminate the algorithm at any time and settle for the best routing solution found so far. This is a very useful feature since reasonably good solutions are obtained very quickly.

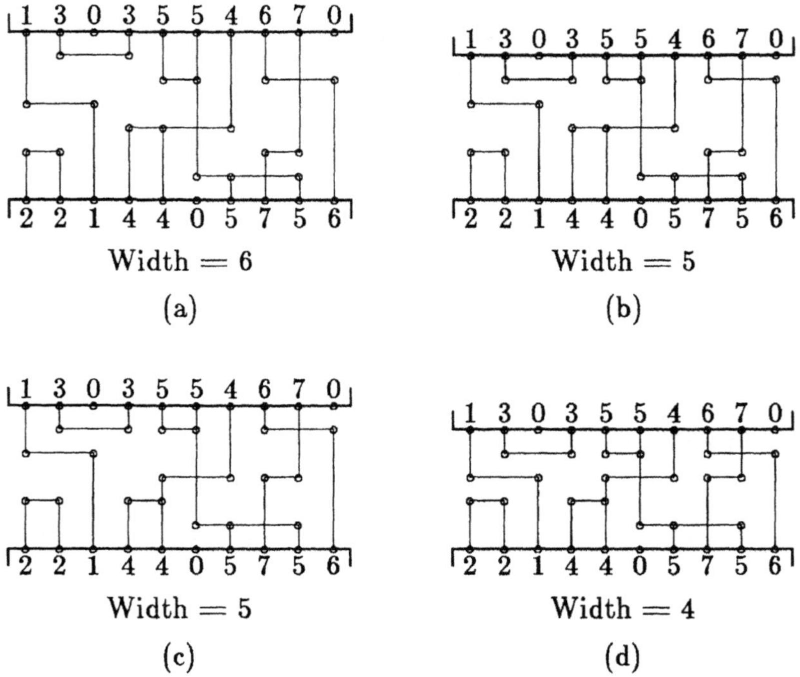

Fig. 4.5 Sequence of solutions examined by SACR.

4.3.1. Solution Space

Given a channel routing problem, we first construct the vertical constraint graph with restricted doglegging G. (See Fig.4.4.) We say that there is a *horizontal constraint* between vertex i and vertex j if the horizontal spans of subnets i and j overlap horizontally. These vertices cannot be placed on the same horizontal track.

Let $\pi = \{V_1, V_2, \ldots, V_w\}$ be a *partition* of the vertex set V into groups. Define G_π, the *merged–graph* induced by π, to be a graph with vertex set $\{ \hat{V}_1, \hat{V}_2, \ldots, \hat{V}_w \}$ where the vertex $\hat{V}_i$ represents the group of vertices V_i in the partition. There is a directed edge from $\hat{V}_i$ to $\hat{V}_j$ in G_π if and only if there exists a directed edge from some $x \in V_i$ to some $y \in V_j$ in G. For example, Fig.4.6(a) shows a partition π, and Fig.4.6(b) shows the corresponding merged–graph G_π.

$\pi = \{V_1, V_2, V_3, V_4, V_5, V_6\}$

$V_1 = \{1\}$ $V_2 = \{2,7\}$

$V_3 = \{3\}$ $V_4 = \{4a,4b\}$

$V_5 = \{5a,6\}$ $V_6 = \{5b,5c\}$

(a)

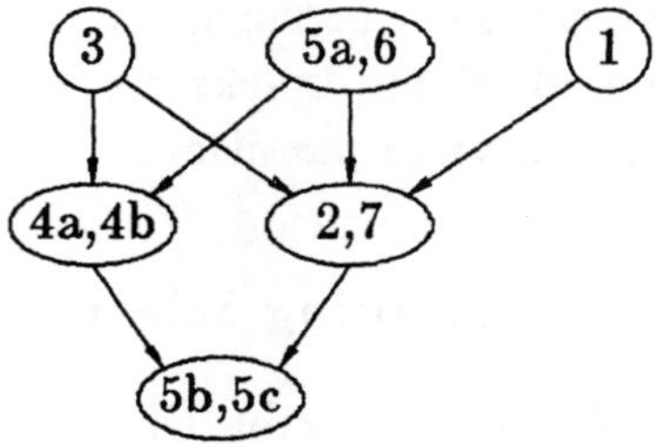

(b)

Fig. 4.6 Solution representation in SACR. (a) The partition π. (b) The merged graph G_π.

A partition π of V is said to be *valid* if (1) G_π is acyclic, and (2) there is no horizontal constraint between any two vertices in the same group. For example, the partition π shown in Fig.4.6(a) is valid since the merged graph G_π (shown in Fig.4.6(b)) is acyclic, and there is no horizontal constraint among subnets in any block in the partition π, namely, between subnets (1) 2 and 7, (2) 4a and 4b, and (3) 5b and 5c.

We now show the equivalence between routing solutions and valid partitions. Given any routing solution that uses w horizontal tracks, we define a partition $\pi = \{V_1, V_2, \ldots, V_w\}$ where V_i consists of all the subnets occupying track i. Clearly, π is a valid partition. For example, the valid partition corresponding to the the routing solution shown in Fig.4.5(a) is exactly that shown in Fig.4.6. Conversely, every valid partition can be mapped into a routing solution as follows: We first label the vertices in G_π topologically [AhHU74]. Then each group V_i of subnets is assigned to track $t(\hat{V}_i)$ where $t(\hat{V}_i)$ is the topological label of $\hat{V}_i$. The routing solution shown in Fig.4.5(a) is obtained from the valid partition shown in Fig.4.6 by topologically labelling the vertices in G_π as follows: $t(V_3)=1$, $t(V_5)=2$, $t(V_1)=3$, $t(V_4)=4$, $t(V_2)=5$, $t(V_6)=6$.

Notice here that w (the number of groups in the partition) is the number of horizontal tracks in the solution. In general, there may be many routing solutions corresponding to one valid partition due to different topological labellings of the vertices in G_π.

However, they are all equivalent in the sense that they all use the same number of tracks w. Thus, the solution space of the channel routing problem is the space of valid partitions and minimizing the number of wiring tracks corresponds to minimizing the number of groups in a valid partition.

4.3.2. Neighboring Solutions

We now describe the notion of neighboring solutions in the solution space. Given a partition π, we define three types of moves to locally modify π:

M1. Two subnets belonging to different groups V_i and V_j are swapped (*i.e.* they change groups).
M2. A subnet is moved from group V_i to group V_j.
M3. A subnet is removed from group V_i to form a new group by itself.

We first show that all the moves are reversible. Clearly, the inverse of a move of type M1 is another move of type M1. The inverse of a move of type M3 is a move of type M2. The inverse of a move of type M2 is a move of either type M2 or M3. Also, given any valid partition π, it is easy to transform π into the trivial valid partition $\{ \{v_1\}, \{v_2\},..., \{v_m\} \}$ by repeatedly applying moves of type M3. By reversing this sequence of moves, it is possible to go from the trivial valid partition to π using only moves of types M2 and M3. Thus, moves of types M2 and M3 are sufficient to ensure that it is possible to go from a given valid partition to any other valid partition.

Two valid partitions π_1 and π_2 are said to be *neighbors* if one can be obtained from the other via one of the three types of moves. Note that the algorithm accepts only valid partitions. For a given valid partition π, we can choose a neighbor of π by repeatedly applying random moves on π until a valid partition is reached. For example, Fig.4.7 shows the valid partitions that correspond to the routing solutions shown in Fig.4.5. In Fig.4.7, the valid partition π_2 is obtained from π_1 by making a move of type M2, moving subnet 3 into the group {5a,6}. Similarly, π_3 is obtained from π_2 after swapping subnets 4a and 7, a move of type M1, and so on.

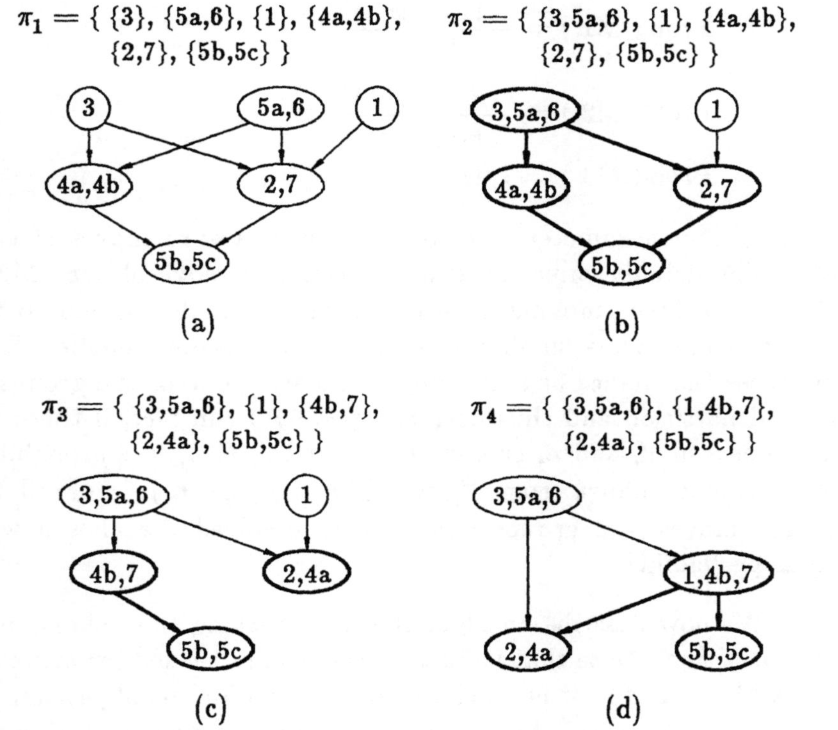

Fig. 4.7 Moving from one solution to another. (a) A solution. (b) After moving net 3 into block {5a,6} (M2). (c) After swapping subnet 4a and net 7 (M1). (d) After moving subnet 1 into the block {4b,7} (M2).

We now consider the effect of these moves on the value of w. It is not difficult to see that a move of type M1 leaves the value of w unchanged, while a move of type M3 increases w by 1. If there is only one subnet in the group V_i, then a move of type M2 will decrease w by one, otherwise w is unchanged. In general, moves of type M2 encourage the reduction of the number of groups, while moves of type M3 allow the algorithm to avoid "getting stuck" at a bad solution. Moves of type M1 are important since they allow rearrangements among solutions with approximately the same cost.

The selection of moves is biased toward a move of type M2, then a move of type M1, then a move of type M3, for the obvious reason that we want to bias toward moves that reduce the value of

w. The following distribution is used:

$$\text{Prob (M1)} = \frac{1}{1+\beta} \cdot \frac{w-1}{w}$$

$$\text{Prob (M2)} = \frac{\beta}{1+\beta} \cdot \frac{w-1}{w}$$

$$\text{Prob (M3)} = 1/w$$

where β is a parameter to control the bias toward moves of type M2. The default value for β is 3. Also if a move of type M2 is chosen, we favor moving a vertex from the smaller group to the larger group. Thus smaller groups tend to become smaller. This increases the chance of eventually eliminating one of the groups in a valid partition and thus decreasing w by 1. In fact, if there are n_i subnets in V_i and n_j subnets in V_j, and $n_i \leq n_j$, the probability of moving a subnet from V_i to V_j is $n_j/(n_i+n_j)$. For all the chosen moves, the groups and subnets involved are chosen with equal probability.

We now describe an algorithm for checking the validity of a new partition. Note that validity check is only needed for moves of types M1 and M2. It is crucial to design an efficient algorithm for validity check since it has to be performed many times during the simulated annealing process. The data structures used are very simple. For each group V_i in π, we use a sorted *subnet-list* of the horizontal span of all the subnets in the group V_i. This, together with the adjacency-list representation [AhHU74] of G, contains all the information we need about G_π. Such information on G_π will be updated after every move. The algorithm for checking validity proceeds as follows:

(1) To check for horizontal consistency when a subnet j is added to a group V_i, we only need to check whether there is any horizontal constraint between subnet j and any other subnet in V_i. Clearly, this can be done by a linear scan of the subnet-list for the group V_i.

(2) To check for vertical consistency, it is sufficient to show that $G_{\pi'}$ is acyclic where π' is the new partition obtained after the move. This can be done in linear time using a depth first search [AhHU74] of the graph $G_{\pi'}$. However, we can reduce the computation time substantially by observing that only a portion of $G_{\pi'}$ need to be examined. If a move introduces a cycle, then that cycle *must* pass

through at least one of the subnet(s) involved in that move. Hence, we only need to perform the depth first search starting at each of the subnet(s) involved. For example, consider the move from π_1 to π_2 in Fig.4.7. We only need to perform the depth first search starting from subnet 3 in group {3,5a,6}. Thus, only part of the merged–graph $G_{\pi'}$ need to be scanned which is indicated by the heavy lines in Fig.4.7(b). Similarly, only three of the vertices in $G_{\pi''}$ are scanned when moving from π_2 to π_3.

4.3.3. Cost Function

Clearly, the objective function to be minimized is the channel width w. However, w is too crude a measure of the quality of intermediate solutions. Instead, for any valid partition π, the following cost function is used:

$$C = w^2 + \lambda_p \cdot p^2 + \lambda_U \cdot U$$

where p is the longest path length of G_π, both λ_p and λ_U are constants, and U will be defined later.

The first term in C is a function of w, which is the objective function to be minimized. To see why w^2 is used instead of w, let us suppose that a move increases the value of w by 1. Then the corresponding increase, Δ_w, would be $2w+1$. This means that there is a higher penalty on solutions with large w's and lower penalty on solutions with small w's. Had we used w instead of w^2, then all moves that increase the channel width by 1 will suffer the same penalty of 1. The reason for the second term is that p is a lower bound on the number of wiring tracks needed for all solutions derived from π by further mergings of subnets. The third term is defined by $U = \sum_{i=1}^{w} u_i^2$, where $u_i = 1 - \frac{1}{l} \sum_{v \in V_i} |v|$, l is the channel length and $|v|$ is the length of the horizontal segment of subnet v. Here, u_i is the fraction of the track i that is unoccupied. Thus, U is a measure of the sparsity of all the tracks in a routing solution. Intuitively, good routing solutions are densely packed, and hence have small U values.

The value of λ_p is used to control the relative importance of w and p. Setting $\lambda_p = 1$ means assigning equal importance to w

and p. Since w is the quantity we want to minimize, it should assume greater importance than p. Thus, λ_p is always set to be less than 1. In general, change in the value of U, Δ_U, is much smaller than those of w and p. In fact, for any move, the maximum value of Δ_U is only about 0.5 which is negligible compared to Δ_w or Δ_p. Thus, the value of λ_U should be large enough so that Δ_U is comparable with the changes in the first two terms but not a great deal larger than that.

Clearly, the choice of suitable values for λ_p and λ_U is important. Experimental results indicate that λ_p=0.5 and λ_U=10 are good choices. These values are used as default values. Also, it was observed that if the value of λ_p is too small (*i.e.* essentially ignoring the longest path), there are considerable fluctuations in the final solution. However, if the value of λ_p is close to 1, it tends to inhibit movements among solutions with the same number of groups, thus decreasing the chance of moving to a better solution. Similarly, if λ_U is too large, there is little freedom for the algorithm to rearrange the groups.

4.3.4. Annealing Schedule

The temperature schedule is of the form: $T_{k+1} = r(k) \cdot T_k$, k=0,1,2,..., where $r(k) = 0.8 + 0.3/(w_k - d + 1)$ with w_k being the minimum width computed by the algorithm up through the kth iteration. The value of $r(k)$ ranges from 0.8 to 0.95. It starts with a value of 0.8 and increases monotonically to a value of 0.95 when $w_k = d+1$. Thus, the rate of decrease in temperature is faster at high temperatures and slower at low temperatures. Such an adaptive schedule works well since it allows more rearrangements among solutions with the same width at low temperatures.

To estimate the value of T_0, we observe that if a move increases w by 1 then the corresponding change in the cost function (denoted by Δ) is $\Delta \simeq 2w+1$ and the probability that this move will be accepted is $e^{-\Delta/T}$. So that there is a reasonable probability of acceptance at high temperatures, we should have $T_0 = O(w)$. Thus, the router starts with T_0 of the form $\lambda_T \cdot w_0$ where w_0 is the channel width of the initial solution. The parameter λ_T is used to compensate for initial solutions that are too far away from an optimal solution.

The channel router can start with any initial valid partition. The default initial solution is a valid partition in which each group V_i consists of all the subnets of net i. In the case that this is not a valid partition, the router would start with the trivial partition in which every subnet forms a group by itself. Clearly, these initial solution are far from optimal.

Let N be the number of neighboring solutions. We now estimate N. There are m possible choices for a move of type M3. The number of moves of type M2 is at most $m(w-1)$ since there are m subnets and each of which can be moved to at most $w-1$ other groups. The number of ways of choosing two subnets for a move of type M1 is given by $\sum_{i<j} m_i m_j$ where $m_k = |V_k|$, $k=1,...,w$. This sum is maximum when all the m_k's are equal. In that case the sum has a value of approximately $m^2/2$. Thus, N is bounded by $m+m(w-1)+m^2/2 = m(m+2w)/2$. Therefore, in the algorithm, N is set to $\lambda_N \cdot m^2$, where λ_N is a constant used to control the number of neighboring solutions.

At each temperature, enough moves are attempted until either there are $c_1 N$ downhill moves or the number of moves attempted exceeds $c_2 N$ where N is the number of neighboring solutions. The termination conditions of the annealing process are: (1) the number of accepted moves is less than 5% of all the moves made at the temperature; or (2) the temperature is low enough. The default values of c_1 and c_2 are 1 and 2, respectively.

Finally, if at any time, a solution using d tracks is reached, where d is the density, the solution is clearly optimal and the annealing process can be immediately terminated.

4.3.5. Fast Approximation Scheme

In order to reduce the running time of the router, a scheme that closely approximates the standard simulated annealing process is introduced. For a given move, we normally first check if the new solution generated is valid or not. If it is, we then compute Δ and then, if Δ is positive, the probability of acceptance or rejection as shown in Fig.4.8(a). Thus, even though checking validity and

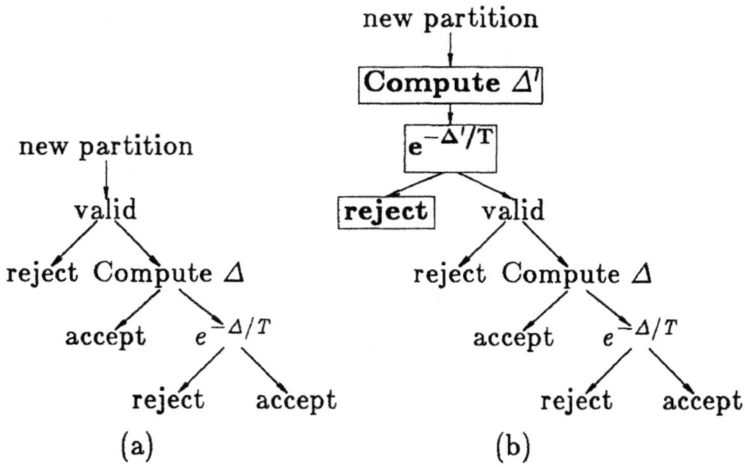

Fig. 4.8 Approximation scheme for simulated annealing.

computing Δ are costly operations, both must be carried out for every move. However, it is easy to see that even if a move of type M3 generates a valid solution, the solution will most likely be rejected since it will lead to a large increase in the cost function. (The value of w will be increased by 1.) Thus, even though most of the moves of type M3 will be rejected, the costly operations of checking validity and computing Δ still have to be performed.

In the approximation scheme, all moves of type M3 are preprocessed as follows: First, we approximate Δ by Δ' then we compute the probability of acceptance or rejection using Δ'. Such preprocessing will eliminate a fraction, α, of the moves of type M3. The other moves will be processed according to the usual procedure. The approximation scheme is shown in Fig.4.8(b). Note that the preprocessing step in the approximation scheme is an approximation of the acceptance/rejection procedure used in the standard scheme (shown in Fig.4.8(a)).

In particular, the time tradeoff is as follows: extra time is spent on computing Δ' for each move of type M3, while for the fraction α of moves rejected on the basis of Δ', we save the time spent on checking validity and computing Δ. Thus, in order for this scheme to be beneficial, we must ensure that Δ' is a good approximation of Δ, that checking validity and computing Δ are computationally more expensive than computing Δ', and that the fraction α is not too small.

In the simulated annealing channel router, Δ' is set to equal $2w+1$ which will closely approximate Δ. Clearly Δ' is easy to compute, while checking validity and computing Δ is time–consuming. The initial experiments indicate that this approximation scheme works very well in the simulated annealing channel routers.

4.4. The Channel Router SACR2

In this section, we describe the extension of SACR to SACR2, a simulated annealing channel router that allows unrestricted doglegging. In SACR, doglegs were automatically created whenever two adjacent subnets occupy different tracks. (See, for example, the dogleg at column 6 between subnets 5a and 5b in Fig.4.5(a).) Thus, doglegs for a net can only occur at columns where there is a pin that belongs to that particular net. In unrestricted doglegging, we allow doglegs be introduced at any column of the channel. In SACR2, we allow doglegs for a net be introduced at any column within its span.

Suppose that for subnet n a dogleg is introduced at column j within its span $[j_1,j_2]$. The dogleg breaks subnet n into two smaller subnets n_1 and n_2. In Fig.4.9(b), an unrestricted dogleg for subnet 3 is introduced at column 5. This breaks subnet 3 into subnets 3a and 3b with horizontal spans [4,5] and [5,6], respectively. Notice that subnet 3a has a pin–connection on its left and a jog–connection on its right while subnet 3b has jog–connection on its left and pin connection on its right.

We can modify the vertical constraint graph to take into account the effect of breaking up a subnet by an unrestricted dogleg as follows: In the vertical constraint graph, G, we replace the vertex corresponding to subnet n by two vertices corresponding to subnets n_1 and n_2. For each subnet x containing the pin t_j, add the directed edges (x,n_1) and (x,n_2) since subnet x must be placed above subnets n_1 an n_2. Similarly, for each subnet y containing

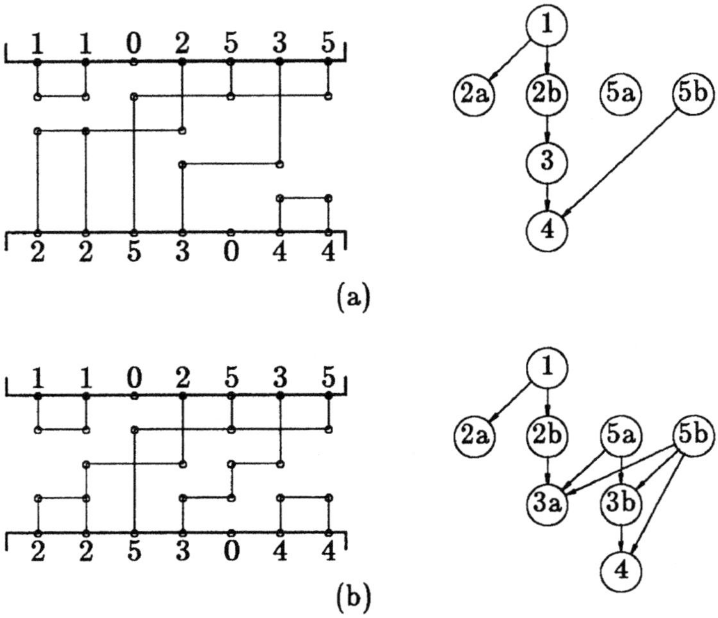

Fig. 4.9 Incorporating unrestricted doglegs. (a) Optimal solution with restricted doglegging. (b) Optimal solution with unrestricted doglegging.

the pin b_j, add the directed edges (n_1,y) and (n_2,y). For the example shown in Fig.4.9(b), we replace vertex 3 by two vertices 3a and 3b in the vertical constraint graph G. Also, we need to introduce two directed edges from subnets 5a and 5b to subnets 1a and 1b. In this particular example, notice that introduction of the unrestricted dogleg breaks the constraint chain (1,2b,3,4) into shorter constraint chains. This leads to a 4 track solution (Fig.4.9(b)) which is better than the optimal solution with restricted doglegging (Fig.4.9(a)). Note that, in general, adding such a dogleg might also introduce cycles in the vertical constraint graph G as illustrated by the example in Fig.4.10.

Suppose a dogleg for a different subnet m is introduced at column j breaking subnet m into two subnets m_1 and m_2. We first perform the transformation described above. Then, if this dogleg is below an existing dogleg in column j, we also add the directed edges (n_1,m_1), (n_1,m_2), (n_2,m_2), and (n_2,m_2). If this

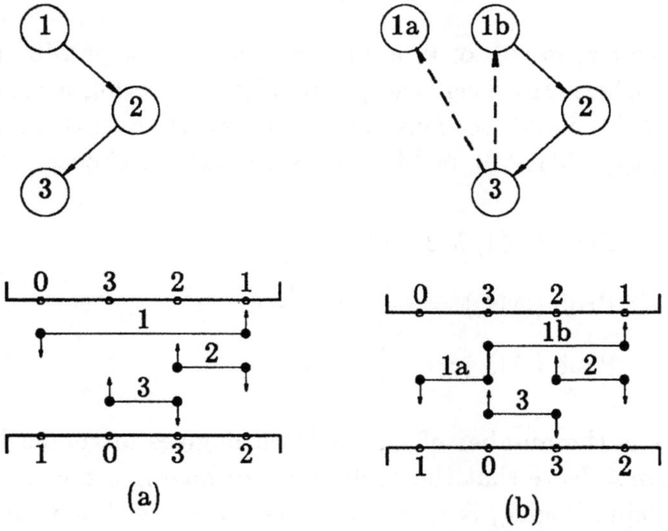

Fig. 4.10 Cycle is introduced by an unrestricted dogleg. (a) No dogleg is used. (b) Introduction of a dogleg.

dogleg is above the existing dogleg in column j, we add the directed edges (m_1,n_1), (m_1,n_2), (m_2,n_1), and (m_2,n_2). Additional doglegs introduced at this column are treated similarly.

In general, suppose that an arbitrary number of unrestricted doglegs has been introduced and the vertical constraint graph has also been appropriately modified. Clearly, the definition of a valid partition π remains unchanged. Furthermore, there is still a correspondence between routing solutions (with unrestricted doglegging) and valid partitions.

Therefore, to incorporate the ability to allow unrestricted doglegging in the simulated-annealing channel router, two additional types of moves are defined:

M4. A subnet n, from group V_i is broken into two subnets n_1 and n_2 by choosing a dogleg randomly within its span. One of these subnets remains in group V_i while the other is moved to another group V_j.

M5. Two subnets, n_1 from group V_i and n_2 from group V_j

that were broken by an unrestricted dogleg are merged together and the merged subnet remains in group V_i.

Clearly, moves of type M4 and moves of type M5 are inverse of each other. However, the probability of making a move of type M4 or M5 should be considerably lower than that of making a move of type M1, M2, or M3. In particular, we choose

$$\text{Prob (M1, M2, M3)} = \frac{w}{w+1}$$
$$\text{Prob (M4)} = \frac{\gamma}{s+\gamma} \cdot \frac{1}{w+1}.$$
$$\text{Prob (M5)} = \frac{s}{s+\gamma} \cdot \frac{1}{w+1}.$$

where s is the number of unrestricted doglegs introduced, and γ is a constant. Note that the probability of having a move of type M4 (introducing dogleg) is monotonic decreasing with s while the probability of having a move of type M5 (merging) is monotonic increasing with s so as to discourage introducing too many unrestricted doglegs.

Since adding an unrestricted dogleg via a move of type M4 might introduce cycles to the merged graph G_π, validity checking must be performed after a move of type M4. (Note that moves of type M5 cannot introduce any cycle in G_π.) The procedure for validity checking is identical to that used for the moves previously described. For horizontal consistency, we verify that the subnet that is moved into group V_j does not conflict horizontally with any of the subnets in group V_j. For vertical consistency, we note that if a cycle is introduced in G_π, then the cycle must pass through either vertex n_1 or n_2. Hence, we perform a depth first search for cycles in G_π starting at these two subnets.

4.5. Experimental Results and Discussion

The channel routers were implemented in Pascal on a PYRAMID computer. The performances of the routers were tested using two types of data: (1) benchmark channel routing problems from [YoKu82]. (YK.3a–YK.5 correspond to ex.3a–ex.5 in [YoKu82] and Deutsch is the well-known Deutsch's difficult problem), (2) randomly generated test problems (R1–R8).

No doglegs

A version of SACR called SACR0 where no doglegging is allowed was also implemented. (SACR0 can be obtained easily by starting with the vertical constraint graph constructed from the nets rather than from the subnets.) The performance of SACR0 was compared with that of other channel routers which do not allow doglegs, namely the Left Edge Algorithm (LEA) of Hashimoto and Stevens [HaSt71] and the "net–merging" router (YK0) of Yoshimura and Kuh [YoKu82] (the version that does not allow doglegs). Table 4.1 shows the number of horizontal tracks used by the various algorithms.[1] In Table 4.1, n is the number of nets, m is the number of subnets, l is the channel length, and d is the channel

Problem Specification					**Channel Width**		
Problem	n	m	l	d	LEA	YK0	**SACR0**
YK.3a	44	74	89	15	18	15	15
YK.3b	47	84	84	17	20	17	17
YK.3c	54	98	103	18	19	18	18
YK.4b	54	128	119	17	23	17	17
YK.5	64	103	128	20	22	20	20
Deutsch	72	218	175	19	30	28	28
R1	19	30	46	10	?	10	10
R2	19	37	47	10	?	10	10
R6	30	56	60	16	?	17	16
R7	26	46	65	14	?	14	14
R8	38	61	63	11	?	11	11

Table 4.1

density. The results indicate that SACR0 obtains optimal solutions for all the test problems. Also, the solutions obtained by SACR0 are considerably better than those obtained by LEA. For Deutsch's difficult problem, SACR0 obtained a solution with 28 horizontal tracks, which is the optimal solution if doglegs are not allowed. Note that there are no results for R3–R5 because these problems have cycles in the vertical constraint graph.

Doglegs

The routers SACR and SACR2 were also compared with two other well–known routers that allows doglegs, the "greedy" router (RF) by [RiFi82] and the dogleg version of the "net–merging" router (YK1) by [YoKu82].[2] The results are summarized in Table 4.2. For all the benchmark problems, the simulated annealing routers perform as well as YK1 and RF. Optimal results were obtained for all except Deutsch's difficult problem which requires 20 tracks, one track beyond optimal. For the randomly generated problems R1–R8, the routers consistently outperform both YK1 and RF. For example, SACR and SACR2 were able to route R6 optimally using 16 tracks whereas an extra track is required by both YK1 and RF. SACR2 produced optimal solutions for all the problems R1–R8 while SACR produced optimal results for all except R5.

The advantage of SACR2 over SACR is illustrated by the results for problem R5. For this problem, SACR uses 16 tracks whereas an optimal solution with 15 tracks was obtained by SACR2.

Computation Time Tradeoffs

The running times of the routers range from a few seconds to a few hours. In particular, the running times for SACR0 are less than 20 minutes for all the test problems. However, the running times for SACR and SACR2 are quite a bit more because the problem size increases quite a bit when doglegs are allowed. For

---

[1] The results for YK0 are obtained from Table III of [YoKu82].

[2] The routers, RF and YK1, have also been implemented by the first and second authors, respectively.

Problem Specification					Channel Width			
Problem	n	m	l	d	RF	YK1	SACR	SACR2
YK.3a	44	74	89	15	15	15	15	15
YK.3b	47	84	84	17	17	17	17	17
YK.3c	54	98	103	18	18	18	18	18
YK.4b	54	128	119	17	17	17	17	17
YK.5	64	103	128	20	20	20	20	20
Deutsch	72	218	175	19	20	20	20	20
R1	19	30	46	10	10	10	10	10
R2	19	37	47	10	10	10	10	10
R3	28	54	60	16	17	16	16	16
R4	28	49	56	16	16	16	16	16
R5	29	51	57	15	16	16	16	15
R6	30	56	60	16	17	17	16	16
R7	26	46	65	14	14	14	14	14
R8	38	61	63	11	11	11	11	11

Table 4.2

example, the running time was about 8 seconds for YK.5, 4 minutes for YK.3c, 1 hour for YK.3b, and 7.5 hours for Deutsch's difficult problem. In general, running time for a particular problem depends on the average number of (sub)nets per horizontal track, and not on the channel length of the problem. This corresponds to the average number of elements in each block. Thus, if there are few nets per block, it is relatively easy to move all the nets from that block to other blocks.

As was pointed out earlier, it is also possible to trade off computation time against the quality of the solution. Since the routers work with complete solutions, it is easy to terminate the annealing process after a given amount of time and settle for the *best solution* found so far. Since the routers typically discover reasonably good solutions very quickly, this usually means that even if computation time is very important, the routers are still competitive with many of the well-known routers. In almost all the test cases, solutions that are only 1 or 2 tracks beyond the optimal are found very early in the annealing process, namely, even at high temperatures.

Discussion

Finally, we discuss the different channel routing models used. Fig.4.11 shows the hierarchy of channel routing models. We note that routers such as LEA, YK1, and SACR0 assume the most restricted model in which no doglegs are allowed. The restricted dogleg model is used in the Dogleg Channel Router [Deut76], YK1, and SACR. Routers such as RF and the hierarchical channel router of [BuPe83] use a more general model that allows unrestricted doglegging and split nets. (A *split net* is a net that is split into more than one horizontal tracks at some column.)

The model that SACR2 assumes lies somewhere in between. It allows unrestricted doglegs but does not allow any "split nets." Under this model, the best known result for Deutsch's difficult problem is 20 tracks. It is not known, currently, whether there is a 19-track solution for this model. Optimal solutions for this problem (with 19 tracks) have been reported in [BuPe83] and [ReSS86] but their channel routers use a more general model that allows unrestricted doglegging and split nets.

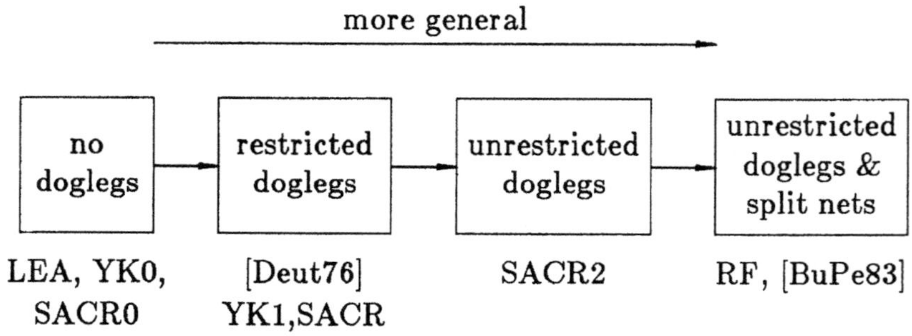

Fig. 4.11 Hierarchy of channel routing models.

Finally, as an indication of the power of simulated annealing, we note that the routers SACR and SACR2 consistently outperform the router RF (see Table 4.2, problems R5 and R6), even though the latter assume a more general model.

4.6. Concluding Remarks

Application of the method of simulated annealing to the channel routing problem is a challenging one because it is not obvious how the method can be applied to this problem. We presented in this chapter two simulated annealing based channel routers SACR and SACR2. The routers are based on a compact representation of channel routing solutions in terms of partitions of the set subnets. This representation leads to a fairly natural notion of neighboring solutions. We also presented a scheme that closely approximate the standard simulated annealing process. This approximation scheme in general saves computation cost. Experimental results show that the routers perform well compare with well-known channel routers such as [YoKu82] and [RiFi82].

CHAPTER 5

PERMUTATION CHANNEL ROUTING

5.1. Introduction

Most of the channel routing algorithms in the literature [Deut76, YoKu82, RiFi82, BuPe83, ReSS85] including the simulated annealing channel routers studied in Chapter 4 assume a model in which the pins on the two sides of the channel are fixed (both physically and logically.) Other models that reflect less stringent physical constraints have also been studied. In [GoCW83], it was assumed that only the relative ordering of the pins is fixed. The pins are allowed to "shift" horizontally. This model captures the possibility of expanding or contracting the dimensions of a cell. In [LeLi83] and [Leon86a], the *discretionary channel routing problem* in which only some of the pins in a given net need to be interconnected was studied. This model captures the possibility of having physically equivalent pins or duplicated pins in a cell. As expected, such flexibilities lead to better routing results, namely, fewer tracks are required for interconnection.

In this chapter, we study the *permutation channel routing problem* (PCRP) in which some of the pins in the channel are interchangeable. Our presentation here is based on [LeLi85] and [LeLi87]. We show that net connections can be significantly improved by allowing such a flexibility. For example, by randomly generating a few groups of interchangeable pins, a savings of 42% was obtained for Deutsch's Difficult Problem. A slightly more restricted version of this model was studied in [KoDr84] where interchangeable pins occur in contiguous blocks. In the model studied in this chapter, there is no restriction on the location of the interchangeable pins.

We show, in this chapter, that the PCRP is NP–complete [GaJo79] for two widely used cost measures – channel density and channel width. We present two different methods of solution for the problem. One is simulated annealing and the other is iterative improvement. Both methods are based on pairwise interchange of pins. Algorithms based on these methods were implemented and tested on various test problems. For the example illustrated in [KoDr84], these algorithms produce a better solution.

This chapter is organized as follows: In Section 5.2, we give the motivation behind the study of the PCRP. Section 5.3 presents the general statement of the PCRP and proofs of NP-completeness. In Sections 5.4 and 5.5, we present two methods of solution for the PCRP, namely, simulated annealing and iterative improvement. The performance of these algorithms are presented in Section 5.6. Finally, some concluding remarks are given in Section 5.7.

5.2. Motivation and Applications

The PCRP is motivated by a wide variety of applications. In the design of fault-tolerant systems, circuits are often duplicated or triplicated so that failure of one of them will not cripple the entire system. Consider the example in Fig.5.1, in which the cells **A** and **A′**, and **B** and **B′**, are duplicates. Clearly, the pins x and x', and the pins y and y', are equivalent. Consequently, the routing

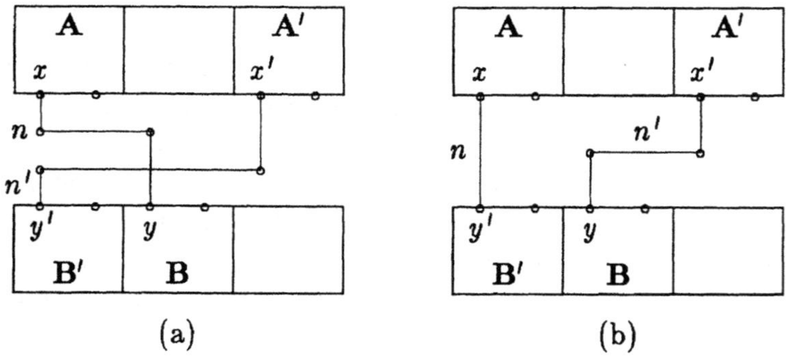

Fig. 5.1 Routing of fault-tolerant circuits. (a) A possible routing. (b) Another possible routing.

problem is that of pairing off the two sets of interchangeable pins $\{x,x'\}$ and $\{y,y'\}$. Fig.5.1(a) and Fig.5.1(b) show the two possible solutions of the associated PCRP.

Programmable logic cells (e.g. ROM's and PLA's) are widely used in VLSI design because of their structural regularity and design flexibility [KoFo84, KoDr84, MeCo80]. Pins within these cells are interchangeable since their geometries are programmable. (See Fig.5.2.) Such situations are conveniently formulated as a PCRP by specifying that all the pins in these cells are interchangeable.

The PCRP arises naturally when there are logically equivalent pins. In Fig.5.3, the different ways to connect the three (equivalent) input pins of an AND–gate to the three pins on the other side of the channel correspond to different permutations of the three input pins. Also, very often, for a circuit with several input (or output) pins, there exists a certain symmetry among the pins such that different placements of the circuit/pins amounts to different permutations of the input (or output) pins. For example,

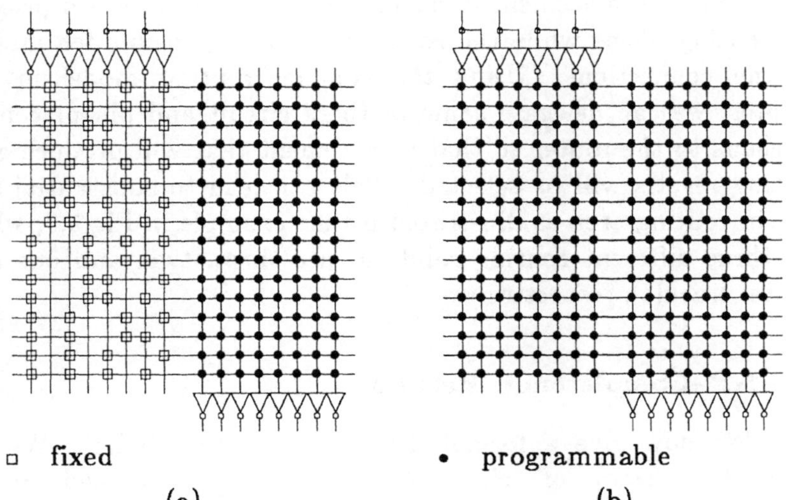

□ fixed • programmable

(a) (b)

Fig. 5.2 Programmable Logic Cells. (a) ROM. (b) PLA.

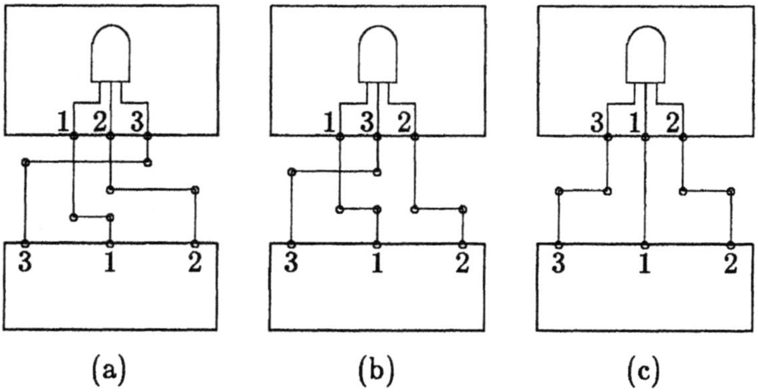

Fig. 5.3 Three possible placements of pins corresponding to three different permutations.

reflection of a three-input gate can be modeled by allowing the leftmost and the rightmost input pins to be interchangeable.

Most of the current design methodologies require that such circuit and pin placement decisions be *completely* specified prior to the routing phase while the router is only responsible for making all the connections. Using the channel routing model in this chapter, we can delegate some of these circuit and pin placement decisions to the router so that a placement that will produce good routing results will be selected. This can lead to substantial savings in routing area as illustrated by the example in Fig.5.3, where the three different routing solution use three, two, and one horizontal track(s), respectively.

5.3. NP–Completeness Results

We now give a formal description of the PCRP. We are given two rows of pins, $T = \{t_s, t_{s+1}, \ldots, t_{l'}\}$ and $B = \{b_1, b_2, \ldots, b_l\}$, along the top and bottom sides of a rectangular wiring channel. Without loss of generality, we can assume $1 \leq s$ and $l' \leq l$. A pin in T (B) is called a *top* (*bottom*) pin. The nets are named 1, 2, ..., n. An additional net named 0 is a fictitious net. If a pin is assigned to net 0, it means that the pin is unconnected. Associated with net i, there are two non–negative integers T_i and

B_i specifying the number of top pins and the number of bottom pins that are to be assigned to net i. An assignment of pins is a function g from $T \cup B$ to $\{0, 1, 2,..., n\}$ so that each pin in $T \cup B$ is assigned to a net and the number of top pins assigned to net i is equal to T_i and the number of bottom pins assigned to net i is equal to B_i. So that we can impose some restrictions on how pins are to be assigned to nets we introduce a partition $\pi_T = \{X_1, X_2, ..., X_i, ...\}$ on T and a partition $\pi_B = \{Y_1, Y_2, ..., Y_j, ...\}$ on B. There is a function f which maps each block X_i is π_T and Y_j is π_B in to a set of multi-sets such that $f(X_i) = \{P_{i1}, P_{i2}, ...\}$ and $f(Y_j) = \{Q_{j1}, Q_{j2}, ...\}$ where each of P_{i1}, P_{i2},..., and Q_{j1}, Q_{j2},... is a multi-set of $\{0, 1, 2,..., n\}$ with $|X_i| = |P_{i1}| = |P_{i2}| = ...$, and $|Y_j| = |Q_{j1}| = |Q_{j2}| = ...$ The assignment function g must be such that g is one-to-one from X_i to P_{i1}, or P_{i2},..., and from Y_j to Q_{j1}, or Q_{j2},.... The meaning of the function f is intuitively quite obvious. A block of pins, such as X_i and Y_j, can be assigned to any one of the multi-sets of nets as specified by P_{i1}, P_{i2},..., and Q_{j1}, Q_{j2},.... Consider the following examples: For

$$f(\{t_1\}) = (\{1\}, \{3\}, \{4\})$$

we mean that pin t_1 can be assigned to net 1, or net 3, or net 4. For

$$f(\{t_1, t_2, t_3\}) = (\{1, 3, 5\})$$

we mean that pins t_1, t_2, t_3 can be assigned to nets 1, 3, 5 in whatever manner we wish. For

$$f(\{t_1, t_2, t_3\}) = (\{1, 1, 4\}, \{0, 2, 5\}, \{1, 3, 5\})$$

we mean that for pins t_1, t_2, t_3, either two of them are assigned to net 1, and the remaining one assigned to net 4; or one of them is unconnected, one of them is assigned to net 2, and the remaining one is assigned to net 5; or they are assigned to nets 1, 3, 5 in whatever manner we wish.

An instance of the PCRP is shown in Fig.5.4(a). A *solution* to the PCRP consists of an assignment together with a corresponding routing. We assume that two layers of interconnect are available for routing, with horizontal wire segments on one layer and vertical wire segments on the other. Fig.5.4(b) and Fig.5.4(c) show two different solutions to the PCRP in Fig.5.4(a). Note, of course,

$$T=\{t_1,t_2,\ldots,t_6\} \qquad B=\{b_1,b_2,\ldots,b_6\}$$

$$\pi_T=\{X_1,X_2,X_3\} \qquad \pi_B=\{Y_1,Y_2,Y_3\}$$

$X_1=\{t_1,t_2,t_4\}$	$f(X_1)=\{\{1,2,3\},\{1,2,4\}\}$	$T_1=2$	$B_1=1$
$X_2=\{t_5\}$	$f(X_2)=\{\{3\},\{4\}\}$	$T_2=1$	$B_2=1$
$X_3=\{t_3,t_6\}$	$f(X_3)=\{1,4\}$	$T_3=1$	$B_3=2$
$Y_1=\{b_2,b_4,b_6\}$	$f(Y_1)=\{\{0,1,4\},\{2,3,4\}\}$	$T_4=2$	$B_4=1$
$Y_2=\{b_1,b_5\}$	$f(Y_2)=\{\{2,3\},\{1,3\}\}$		
$Y_3=\{b_3\}$	$f(Y_3)=\{\{0\},\{3\}\}$		

(a)

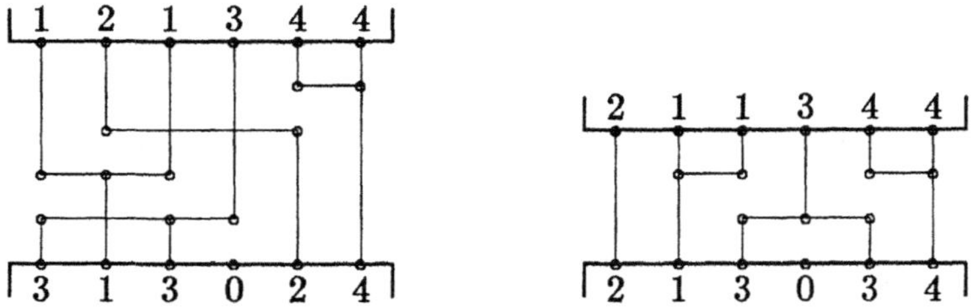

(b) (c)

Fig. 5.4 Permutation channel routing. (a) A permutation channel routing problem. (b) A solution with 4 tracks. (c) A solution with 2 tracks.

that once an assignment, g, is chosen, the PCRP is reduced to a standard channel routing problem.

We define some terminologies. Given an assignment, g, let N_k denote the set of all the pins assigned to net k, and l_k and r_k denote the positions of the leftmost and rightmost pin in N_k, respectively. The *span* of N_k denoted $s(N_k)$, is defined to be $[l_k,r_k]$. The *local density* at column i, denoted $d_i(g)$, is the number of nets whose spans intersect column i. (Nets with span $[i,i]$ are ignored in this definition.) The *density vector* $D(g)$ is defined by $D(g) = (d_1,d_2,\ldots,d_l)$. The *density* $d(g)$, is the maximum local density. For a given solution to the PCRP, the width w is the number of horizontal tracks used in the solution.

Recall that in channel routing, the density is a fixed constant while the width is used as a measure of the quality of the solution. In the PCRP, the density is no longer a constant, rather, it is a function of the assignment. Thus, for the PCRP, we define two measures of quality – the width, w, and the density, d. For example, the solution in Fig.5.4(b) has $d=3$ and $w=4$, while that in Fig.5.4(c) has $w=d=2$.

The problem of finding a solution to the PCRP that minimizes w is clearly NP–complete since the PCRP is a generalization of the channel routing problem which is a well–known NP–complete problem [Szym85, GaJo79]. On the other hand, d is a very good approximation of w since (1) [BaBL84] has shown that $d \leq w \leq 2d + O(\phi)$ where ϕ is the *flux*, and (2) for most practical problems, existing routers produce solutions in which the width w is typically d or very near to d.

Note that $d(g)$ depends only on the assignment g. Let $d_{\min}$ and $d_{\max}$ be the minimum and maximum values of the density for all possible assignments. Hence, it is natural to ask if there are polynomial time algorithms for the problem of determining an assignment that minimizes the density. It should be pointed out here that this might appear to be a simpler problem than that of finding an optimal solution since, for the assignment problem, we only want to determine an assignment – the routing need not be specified. However, it turns out that the assignment problem is also NP–complete and so it is highly unlikely that efficient algorithms can be found for the optimal assignment problem for the general PCRP. We first state two versions of the assignment problem.

Minimum Density Assignment Problem (MDAP)
Instance. An instance of the PCRP and an integer $k \leq n$.
Question. Is there an assignment g with $d(g) \leq k$?

Minimum Width Assignment Problem (MWAP)
Instance. An instance of the PCRP and an integer $w \leq n$.
Question. Is there an assignment that requires w or fewer tracks?

Theorem 5.1. The MDAP is NP–complete.

Proof. The MDAP is clearly in NP, since given any assignment g, it is trivial to check if $d(g) \leq k$. To complete the proof, we shall show a reduction from the Minimum Cut Linear Arrangement Problem (MCLAP) to the MDAP. We first give a formal statement of MCLAP.

Minimum Cut Linear Arrangement Problem (MCLAP)
Instance. A graph (V, E), a positive integer d.
Question. Is there a one–to–one function ϕ: $V \rightarrow \{1, 2, ..., |V|\}$ (a linear arrangement of the vertices in V) such that for all i, $1 < i < |V|$

$$|\{\{u, w\} \in E: \phi(u) \leq i < \phi(w)\}| \leq d?$$

Given an instance of the MCLAP, we construct an instance of the MDAP as follows: For a graph $G = (V, E)$. Let m denote the number of vertices and n denote the number of edges in G. (That is, $m = |V|$, $n = |E|$.) Let the vertices be denoted $v_1, v_2, ..., v_m$. Let $\delta(v_i)$ denote the degree of v_i. Let δ denote the maximum degree of the vertices in G. Let Δ denote $\delta + 2$. Consider a permutation channel routing problem in which $T = \{t_{11}, t_{12}, ..., t_{1\Delta}, t_{21}, t_{22}, ..., t_{2\Delta}, ..., t_{m1}, t_{m2}, ..., t_{m\Delta}\}$, $\pi_T = \{\{t_{11}, t_{12}, ..., t_{1\Delta}\}, \{t_{21}, t_{22}, ..., t_{2\Delta}\}, ..., \{t_{m1}, t_{m2}, ..., t_{m\Delta}\}\}$, and there is no pin in the bottom of the channel. Let there be n two–terminal nets one corresponds to each edge in E. Specifically, the two–terminal net corresponding to the edge $\{u, w\}$ in E will be named e_{uw}. Let there be m two–terminal nets one corresponds to each vertex in V. Specifically, the two–terminal net corresponding to the vertex v_i in V will be named v_i. The function f is defined as followed: For each block $\{t_{k1}, t_{k2}, ..., t_{k\Delta}\}$ in π_T, $k = 1, 2, ..., m$,

$$f(\{t_{k1}, t_{k2}, ..., t_{k\Delta}\}) =$$
$$\{\{e_{uv_1} | u \neq v_1, u \in V\} \cup \{v_1, v_1\} \cup \{\Delta - \delta(v_1) \ \ 0's\},$$
$$\{e_{uv_2} | u \neq v_2, u \in V\} \cup \{v_2, v_2\} \cup \{\Delta - \delta(v_2) \ \ 0's\},$$
$$\cdots\cdots\cdots$$
$$\{e_{uv_m} | u \neq v_m, u \in V\} \cup \{v_m, v_m\} \cup \{\Delta - \delta(v_m) \ \ 0's\}\}$$

We now claim that there is a one–to–one function ϕ from V to $\{1, 2, ..., |V|\}$ such that for all i, $1 < i < |V|$, $|\{\{u, w\} \in E: \phi(u) \leq i < \phi(w)\}| \leq d$ if and only if there is an assignment function g such that the channel density is less than or equal to d.

Let us assume that there is such an assignment function g. Note that the introduction of the two–terminal nets $v_1, v_2, \ldots, v_m$ is a trick that forces the function g to assign the m different blocks of pins in π_T, $\{t_{11}, t_{12}, \ldots, t_{1\Delta}\}$, $\{t_{21}, t_{22}, \ldots, t_{2\Delta}\}, \ldots, \{t_{m1}, t_{m2}, \ldots, t_{m\Delta}\}$, to the m different multi–sets of nets $\{e_{uv_1} | u \neq v_1, u \in V\} \cup \{v_1, v_1\} \cup \{\Delta - \delta(v_1) \text{ 0's}\}$, $\{e_{uv_2} | u \neq v_2, u \in V\} \cup \{v_2, v_2\} \cup \{\Delta - \delta(v_2) \text{ 0's}\}, \ldots, \{e_{uv_m} | u \neq v_m, u \in V\} \cup \{v_m, v_m\} \cup \{\Delta - \delta(v_m) \text{ 0's}\}$. Consider the block of pins $\{t_{k1}, t_{k2}, \ldots, t_{k\Delta}\}$, $k = 1, 2, \ldots, m$. The assignment function g must be such that

$$g(t_{kl}) = e_{uv_i} \text{ or } v_i \text{ or } 0$$

for $l = 1, 2, \ldots, \Delta$, $u \neq v_i$, $u \in V$. It is then obvious that the linear arrangement

$$\phi(v_i) = k$$

is a solution to the MCLAP.

Conversely, suppose that there is a function ϕ which is a solution to the MCLAP. It is obvious that for $k = 1, 2, \ldots, m$ if

$$\phi(v_i) = k$$

we shall define the assignment function g to be such that

$$g(t_{kl}) = e_{uv_i} \text{ or } v_i \text{ or } 0$$

for $l = 1, 2, \ldots, \Delta$, $u \neq v_i$, $u \in V$. We only need to be careful in assigning the pins in the block $\{t_{k1}, t_{k2}, \ldots, t_{k\Delta}\}$ to the nets in the multi–set $\{e_{uv_i} | u \neq v_i, u \in V\} \cup \{v_i, v_i\} \cup \{\Delta - \delta(v_i) \text{ 0's}\}$. We leave it to the reader to confirm that assigning pins in the "left–hand side" of the block to two–terminal nets whose other terminals are at pins that are to the left of the block, assigning two adjacent pins in the "middle" of the block to the two–terminal net v_i, and assigning pins in the "right–hand side" of the block to two–terminal nets whose other terminals are at pins that are to the right of the block as depicted in Fig.5.5 is an assignment that will work. □

Corollary 5.1. The MWAP is NP–complete.
Proof. In the proof of Theorem 1, we use a one–sided channel. It is well–known that for a one–sided channel, minimum channel width is equal to the channel density. □

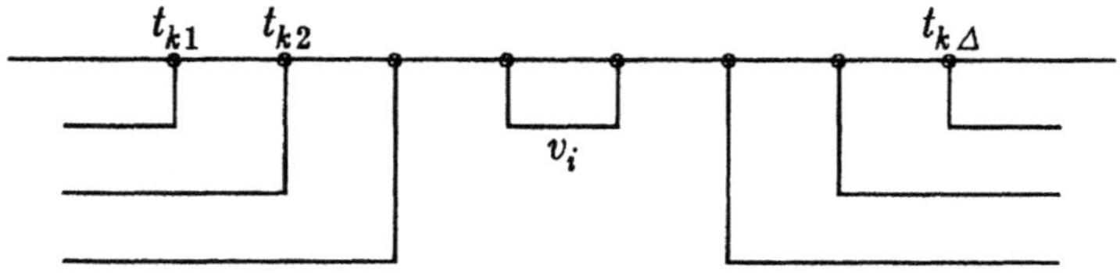

Fig. 5.5 The assignment function g.

5.4. First Method – Simulated Annealing

We shall present in Sections 5.4 and 5.5 assignment algorithms for the PCRP. Our presentation is for a special case in which corresponding to each block of pins there is only one multiset of nets. That is, $f(X_i) = \{P_{i1}\}$, and $f(Y_j) = \{Q_{j1}\}$. In other words, the pins in X_i and Y_j are interchangeable pins for the nets in P_{i1} and Q_{j1}. In this section, we describe an assignment algorithm that employs the method of simulated annealing. The solution space consists of all possible assignments of the interchangeable pins. We now describe the definition of neighboring solutions, the cost function, and the annealing schedule used in the algorithm.

5.4.1. Neighboring Solutions

For the PCRP, there is a natural definition of neighboring solutions. Let g be an assignment and let p_1 and p_2 be two interchangeable pins assigned to two distinct nets k_1 and k_2, respectively. A *move* interchanges the assignments for pins p_1 and p_2 to get to a new assignment g'. We shall denote such a move by $EXCHANGE(p_1,p_2)$. Define the *neighbors* of g to be those assignments that are reachable from g in one move.

For each group of m interchangeable pins, there are $m(m-1)/2$ possible choices of an interchange pair. Thus the total number of interchange pairs is $\sum_{k=1}^{p} s_k(s_k-1)/2 + \sum_{j=1}^{q} r_j(r_j-1)/2$ where $|X_k|=s_k$ and $|Y_j|=r_j$. In the worst case, there are $O(ls)$ possible interchange pairs, where l is the channel length, and s is the size of the largest group of interchangeable pins. Thus, for any assignment g, there are $N=O(ls)$ neighbors and this number is a fixed constant that is independent of g. In practice, s is usually small, and hence it will be safe to assume that $N=O(l)$.

To compute the density vector $D(g)$, after a move, we note that $EXCHANGE(p_1,p_2)$ will affect at most two nets, namely, nets k_1 and k_2, and will leave all the other nets unchanged. Thus, a straight–forward way to compute the new d_i's after an *EXCHANGE* operation is to evaluate the changes in local densities due to the changes of pin positions of the two nets k_1 and k_2. Let l_k' (r_k') denote the position of the leftmost (rightmost) pin assigned to the net k after the *EXCHANGE* operation. Then the changes in local densities corresponding to the new pin positions of net k can be easily computed using the procedure shown in Fig.5.6 where $+[i_1,i_2]$ ($-[i_1,i_2]$) means that the local density of every column in the interval $[i_1,i_2]$ is increased (decreased) by one. Hence, the new d_i's can be computed by applying the procedure in Fig.5.6 to nets k_1 and k_2.

The disadvantage of this simple straight–forward method is that some of these computations might be redundant. Consider the move from g to g' shown in Fig.5.7. Clearly, $D(g)=D(g')$ and no computations are necessary to obtain $D(g')$ from $D(g)$. However, if the procedure in Fig.5.6 is applied to nets 1 and 2, we would perform the (redundant) updates $-[1,1]$ and $+[1,1]$. In general, the amount of redundant computations can be as large as $O(l)$. Since the methods involves many (thousands of these) *EXCHANGE* operations, it is crucial that all redundant

begin
  if $(l_k'<l_k)$ **then** $+[l_k', l_k-1]$;
  if $(l_k'>l_k)$ **then** $-[l_k, l_k'-1]$;
  if $(r_k'>r_k)$ **then** $+[r_k+1, r_k']$;
  if $(r_k'<r_k)$ **then** $-[r_k'+1, r_k]$;
end

Fig. 5.6 Changes in local densities corresponding to the new pin positions of net k.

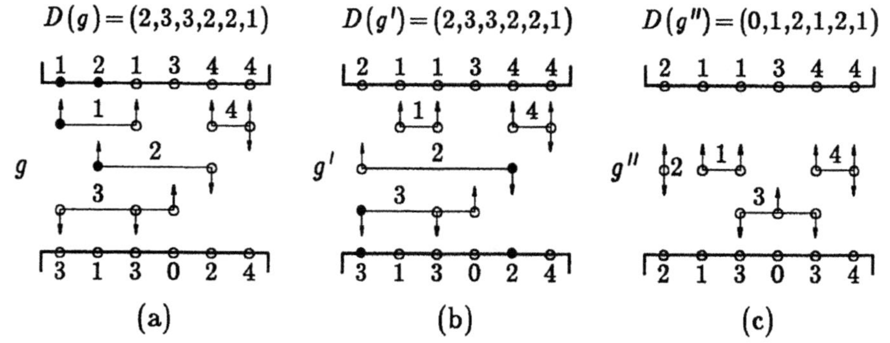

Fig. 5.7 Moving from g to g' to g''. (a) An assignment g. (b) After $EXCHANGE(t_1,t_2)$. (c) After $EXCHANGE(b_1,b_5)$.

computations be avoided. Let Δ_D denote the change in $D(g)$ in moving from g to g' that avoids all redundant computations. We require a fast characterization of Δ_D. Indeed, this characterization can be obtained in constant time using very simple data structures.

The idea is to examine all possible configurations of the pins in nets k_1 and k_2. We then divide these configurations into classes and for each class give the corresponding Δ_D. This can be done using only a small number of simple comparisons. We outline the method here. First assign a label $M(p)$ to each pin p in the channel as follows: A pin has label U (unused) if it is assigned to net 0, L (leftmost) if it is a leftmost pin of a net, R (rightmost) if it is a rightmost pin of a net and I (interior) if it is an interior pin of some net. Namely, for pin t_i,

$$M(t_i) = \begin{cases} U & \text{if } g(t_i)=0 \\ L & \text{if } g(t_i)\neq 0 \text{ and } i=l(k) \\ R & \text{if } g(t_i)\neq 0 \text{ and } i=r(k) \\ I & \textit{otherwise} \end{cases}$$

Note that bottom pins b_i are assigned labels $M(b_i)$ too.

Without loss of generality, we shall assume that p_1 is to the left of p_2 and that $k_1=1$ and $k_2=2$. The labels divide the configurations into different types according to the labels $M(p_1)$ $M(p_2)$.

Thus there are sixteen such types LL, LU, LI, LR, ... ,RR. However, configurations of a given type may have different Δ_D. Hence, within a given type, the configurations are further sub-divided into classes so that all configurations in a given class have the same Δ_D.

The method of sub-division is based on the relative positions of the pins in nets 1 and 2. In particular, this method involves only a few comparisons to uniquely determine the class to which a given configuration belongs. We illustrate this sub-division using a few examples. First, we need to define some terminologies. The position of a pin p_1 is denoted by x_1 if it is an interior pin of net 1, and by u_1 if it is unused. Also, let s_1 and v_1 denote the position of the second leftmost and second rightmost pins in net 1, respectively. The counterparts, x_2, u_2, s_2, and v_2 are defined similarly.

For example, the configurations of type LL are sub-divided into three classes, namely, Classes 1, 2, and 3 as shown in Fig.5.8. (Note that only the spans of nets 1 and 2 are shown in Fig.5.8 since all the other nets are unaltered.) Class 1 consists of all configurations in which $(l_2 \leq r_1)$ and $(s_1 \geq l_2)$ with corresponding $\Delta_D = \phi$. Clearly, $D(g') = D(g)$ and notice that the redundant computations in the interval $[l_1, l_1'-1]$ are avoided. Class 2 consists of all configurations in which $(l_2 \leq r_1)$ and $(s_1 < l_2)$ with corresponding $\Delta_D = +[l_1', l_2-1]$. Again, redundant computations are avoided in the interval $[l_1, l_1'-1]$. Finally, Class 3 consists of all configurations in which $(l_2 > r_1)$ and for this class $\Delta_D = \{ +[l_1', l_2-1], +[r_1+1, r_1'] \}$. As another example, all configurations of type RL belong to the same class, namely, Class 42, and for these configurations, $\Delta_D = \{ +[l_2', l_2-1], +[r_1+1, r_1'] \}$ as shown in Fig.5.9.

The sub-division of the configurations of all the other types can also be done similarly. Based on this sub-division, there is a total of 48 distinct classes, namely, Class 1–48. (See [Leon86a] for an exhaustive list of the different classes together with the conditions that characterize each class and the corresponding Δ_D's.)

Notice that Δ_D consists of at most two interval updates with no redundant computations. Also, there are only 8 different interval updates, namely, $-[l_1, l_1'-1]$, $-[l_2, l_1'-1]$, $-[r_2+1, r_2]$ $-[r_2'+1, r_1]$, $+[l_1', l_2-1]$, $+[l_2', l_2-1]$, $+[r_1+1, r_1']$, $+[r_1+1, r_2']$. This information can be easily encoded and stored in a table for fast table look-up.

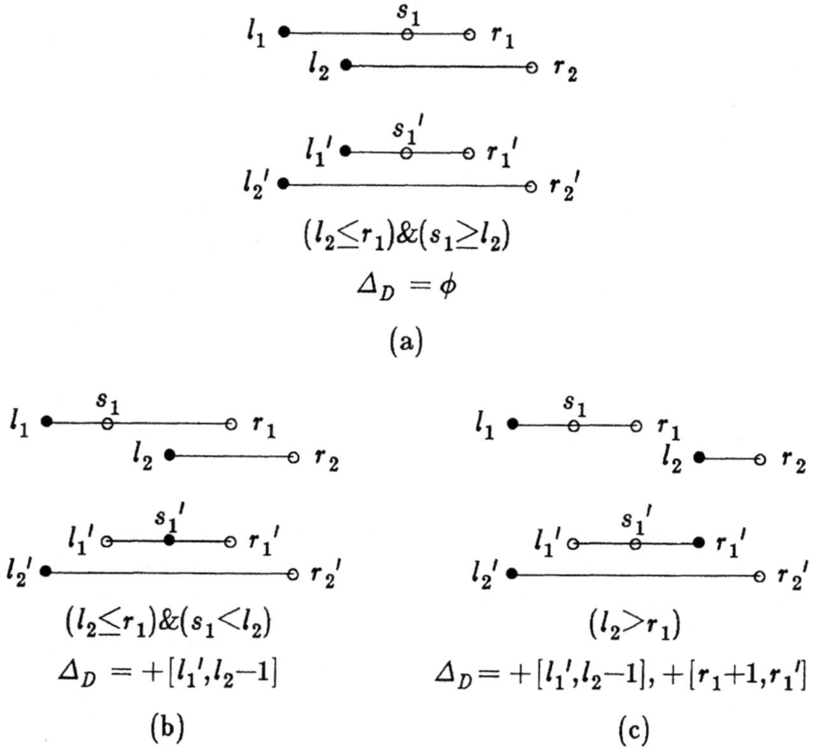

Fig. 5.8 Configurations of Type LL. (a) Class 1. (b) Class 2. (c) Class 3.

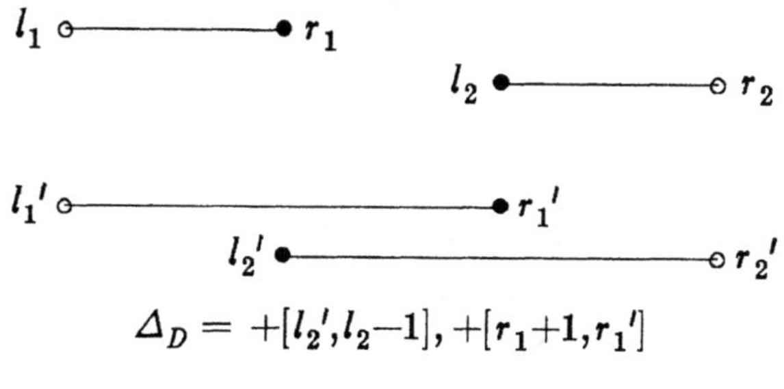

Fig. 5.9 Configurations of Type RL – Class 42.

Thus, for any move $EXCHANGE(p_i, p_2)$, Δ_D can be computed in constant time: First determine the class to which

EXCHANGE(p_1,p_2) belongs using a constant number of comparisons, then do a table look–up for Δ_D.

The data structures used to support all these operations are very simple. For each net k, maintain a sorted list of pins assigned to the net. *EXCHANGE*(p_1,p_2) involves removing the pin p_1 from the list for net k_1 and inserting it into the list for net k_2 and conversely for pin p_2. The labels for the pins are updated accordingly. If there are m pins in a list, these operations can be done in time $O(\log m)$ if the list is maintained as a height–balanced tree [AhHU74] or in time $O(m)$ if it is just a sorted linked list. Typically, m is very small and hence a simple sorted linked list will be sufficient.

5.4.2. Cost Function.

Recall that the objective is to minimize the density d. However, d is too crude to be used as the cost function since it rarely changes and is insensitive to small local changes that do not alter the density. In the simulated annealing algorithm, the following cost function is used:

$$C(g) = \lambda d^2(g) + \frac{1}{l}\sum_{i=1}^{l} d_i^2(g).$$

The first term in C is a function of the density d and the second term is the average value of the squares of the local densities. The reason for the first term is clear since the goal is to minimize $d(g)$. The second term serves to favor solutions that have smaller local densities among solutions with the same density. Since $d_i \leq d$ for each i, it is clear that the first term is the dominant term (provided that $\lambda > 1/l$, a condition that is true in all practical cases). Thus, changes in d will play a major role while changes in the d_i's will be noticeable only if d is unchanged.

The reasons for the squaring are twofold. First, solutions with high densities will suffer higher penalties than those with low densities. If the density increases from d to $d+1$, the increase in cost will be $2d+1$. Thus, the penalty suffered is directly proportional to the density; higher when d is large. Also, even among solutions with the same density, columns with high local densities will suffer higher penalties than those with low local densities.

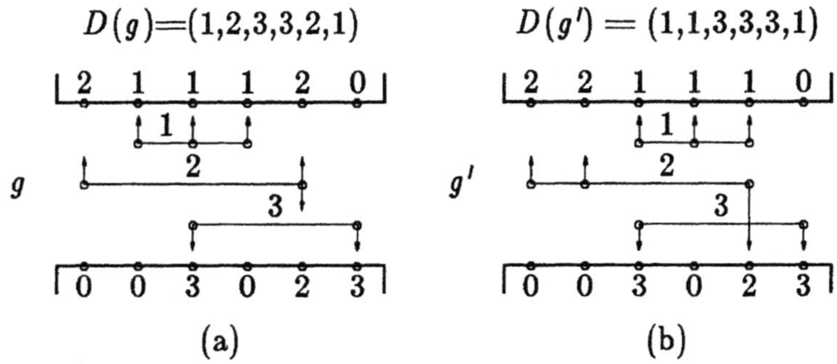

Fig. 5.10 $D(g)$ and $D(g')$ are incomparable. (a) An assignment g. (b) After $EXCHANGE(t_2,t_5)$.

Second, squaring the local densities will avoid cancellation of increments and decrements of local densities at different columns. For the example in Fig.5.10, change in the cost function would have been cancelled out if d and d_i had been used in C instead of d^2 and d_i^2 since the decrease at column 2 is more than offset by the increase at column 5.

The value of λ controls the relative importance of the two terms in the cost function. Setting λ very large corresponds to using the density as the cost and ignoring the fluctuations in local densities among all solutions with the same density. This is undesirable since it defeats the purpose of including the second term, and we also know that density d alone is not an effective cost function. On the other hand, setting λ too small corresponds to relying on the average of the d_i^2's rather than on d^2 as the primary measure. To obtain some reasonable lower bounds on the value of λ, we compute the mean value of the second term for various distributions of the local densities d_i's. For a given density d, if we assume that the local densities are uniformly distributed over $[0,1,\ldots,d]$, then the expected value of the second term is approximately $d^2/3$. If a binomial distribution is assumed, it would be $d^2/4$. Thus for λ to be meaningful, we should choose $\lambda \geq 1/3$. Experimental results show that λ=0.5 produces good results.

To determine whether $C(g') \leq C(g)$, it is not sufficient to just compute Δ_D, we must also use Δ_D to compute the *actual* change in the cost function $C(g)$. However, the neighborhood characterization gives a fast computation of Δ_D and also ensures that redundant computations can be avoided. Thus, the change in the cost function C can be computed very efficiently.

5.4.3. Annealing Schedule

The temperature schedule is defined as follows: $T_k = r \cdot T_{k-1}$, $k=1,2,...$. To estimate reasonable values of the initial temperature, T_0, we consider the following situation. Suppose that an interchange increases the density from d to $d+1$. Then the change in the function C will be $\delta = \lambda(2d+1) + o(d)$, and the probability of making this move is $e^{-\delta/T}$. Thus, for there to be any reasonable probability of making this move at the initial "high" temperature T_0, we must have $T_0 = O(\lambda d)$. Thus, the initial temperature T_0 is chosen to be of the form $\lambda_T \cdot \lambda d$, where λ_T is used to control the initial temperature. The default value of λ_T is 0.5. The value of r varies from 0.8 to 0.95. Very rarely do we need to slow down the annealing process to $r=0.95$. The default value of r is 0.8.

At each temperature, enough moves are attempted until either $c_1 N$ moves are accepted or $c_2 N$ moves are attempted where N is the number of neighbors. The default values for c_1 and c_2 are 2 and 8, respectively.

For the PCRP, as $d(g)$ approaches the minimum density, it is progressively more difficult to get to a configuration with a lower density; in other words, the algorithm needs to move around longer among solutions with the same density before it discovers a neighbor with a lower density. It is not at all surprising to go through two or three iterations at low temperature before encountering a solution that has a lower density. Of course, the algorithm should keep track of the best solution found so far and also the minimum density after each iteration.

The annealing process is terminated if there has been no change in the $d(g)$ for c_3 (default $c_3=6$) successive temperatures or if the number of downhill moves is less than 5% of all the accepted moves.

5.5. Second Method – Iterative Improvement

We now describe the iterative improvement approach to the PCRP. There are two versions based on two different cost functions. The first is a very fast version which produces reasonable results; the second is not as fast as the first (but still very fast) and produces better results. We first give an overview of the iterative improvement scheme.

5.5.1. The Iterative Improvement Scheme

In the method of iterative improvement, a cost function is chosen as a measure of the quality of a solution. Starting with an initial assignment g in the solution space, we examine the neighboring solutions one after another. If there is a neighbor g' with a lower cost, the appropriate *EXCHANGE* operation is performed and g' becomes the current solution. The algorithm then looks for further improvement starting at g' and so on. This process terminates when no further improvement can be found, i.e. the current assignment is a local (or global) optimum.

To improve upon this basic scheme, the algorithms also try to avoid remaining stuck at a local optimum, say g, by allowing "ϵ—moves." An *ϵ—move* is a move in which g and g' have the same cost. Note that ϵ—moves will enable the algorithms to move from a local minimum g (namely, there there is no neighbor with a lower cost), to a new assignment g'. If after some number of such ϵ—moves, it arrives at a solution that has a neighbor with a lower cost, then the algorithm has successfully moved out of the local minimum g.

To incorporate these into the algorithms, we shall scan the neighbors as follows: The possible moves are ordered $1,2,...,N$. The neighbors of g are scanned one-by-one by scanning the corresponding moves. Instead of starting with move 1 all the time, the algorithms remember the move last scanned and proceed from there. The first neighbor with a lower cost will be accepted as the new solution. The algorithms also keep track of the first ϵ—move encountered. If there is no neighbor with a lower cost, this ϵ—move will be chosen. The algorithms terminate when they have made more than a certain number, N_ϵ, of consecutive ϵ—moves without finding any further improvement.

We describe below two different algorithms based on two different cost functions D and C. The corresponding algorithms for the cost functions D and C will be referred to as Version–D and Version–C, respectively.

5.5.2. Version–D

For Version–D, the cost function is the density vector $D(g) = (d_1, d_2, ..., d_l)$. For two assignments g and g', we say that $D(g') \leq D(g)$ if and only if at every column i, $d_i(g') \leq d_i(g)$. We also say that $D(g') < D(g)$ if $D(g') \leq D(g)$ and $D(g') \neq D(g)$.

In Fig.5.7, the first move from g to g' is an ϵ—move while the second move results with a solution with lower cost *i.e.*, $D(g') > D(g'')$. Note that both g' and g'' have the same density. Thus, the cost function $D(g)$ enables the algorithm to distinguish between solutions with the same density. It is also possible that neither $D(g') \leq D(g)$ nor $D(g) \leq D(g')$ as in the case in Fig.5.10. In this case, $D(g)$ and $D(g')$ are said to be *incomparable.*

The major advantage of using the cost function $D(g)$ is that each neighbor can be processed in *constant time.* We do not need to compute the cost function $D(g)$ explicitly since we only need to know whether $D(g) \leq D(g')$, and this can be done in constant time by computing Δ_D using the neighborhood characterization. Thus, no numerical computation is necessary and the algorithm is *extremely fast.* The disadvantage of using this cost function is that the algorithm does not make incomparable moves and this may cause the algorithm to miss some potentially good moves.

5.5.3. Version– C

For Version–C, the cost function $C(g)$ is the same as the one used in the simulated anneaing algorithm. Intuitively, this cost function uses the density $d(g)$ as the primary measure and the $d_i(g)$'s as the secondary measure. Thus, it is more sensitive to local changes. The advantage of Version–C over Version–D is that it is less restrictive and therefore produces *better results.* In fact, any downhill neighbor with respect to the cost function $D(g)$ will also be a downhill neighbor with respect to $C(g)$. However, some downhill neighbors with respect to $C(g)$ may be incomparable with respect to $D(g)$. Thus, $C(g)$ defines more downhill neighbors and tends to allow more "good" moves. Furthermore, this cost function is extremely sensitive to small local changes. Version–C is not as fast as Version–D since change in the cost function needs to be computed explicitly. However, it is still a fast and practical algorithm.

5.5.4. Choice of Initial Solution

Since the results for the iterative improvement algorithms are dependent on the choice of an initial solution, a number of different initial solutions were used for each algorithm and the best solution is kept. The initial solutions include random solutions as well as solutions that are "far" from optimal (clearly "bad" solutions). Experimental results indicate that both versions perform well on both types of initial solutions. In fact, for most of the randomly generated test problems, no noticeable difference was observed in the results from both types of initial solutions.

5.6. Experimental Results

The algorithms described in Sections 5.4 and 5.5 were implemented in Pascal on a Cyber 175 computer. All three algorithms were tested on various types of test problems.

Table 5.1 shows their performance on three difficult sample problems (D1–D3) that have been specially constructed to test the algorithms. The minimum density for each of these problems is 2. These problems are difficult because most of the assignments yield a larger density. The algorithms were able to find an optimal assignment in each case.

Table 5.1 also summarizes the results obtained for some randomly generated PCRP's. For each problem (R1–R8), the groups of interchangeable pins are generated randomly.

The algorithms were also tested on some channel routing

Test Problem	l	n	d_{max}	Iterative Improvement		Simulated Annealing
				Version–D	Version–C	
D1	25	15	13	2*	2*	2*
D2	40	20	18	2*	2*	2*
D3	40	20	10	2*	2*	2*
R1	48	20	20	9	9	9
R2	50	20	20	9	9	8
R3	59	30	29	12	11	10
R4	77	30	30	12	11	10
R5	95	40	40	18	17	17
R6	103	40	40	22	21	20
R7	124	50	49	24	23	22
R8	128	50	50	24	22	21

* (minimum channel density.)

Table 5.1

examples[1] in the literature [YoKu82, Deut76]. In order to convert these standard channel routing problems to PCRP's, some *small* groups of interchangeable pins are generated for each of these problems. Table 5.2 shows the performance of the algorithms on these problems. As Table 5.2 indicates, substantial saving in the number of tracks can be obtained by allowing interchangeable pins. In particular, for Deutsch's Difficult Problem [Deut76], the number of tracks decreases from 19 to 11, a saving of 42%. For YK.3a from [YoKu82], the number of tracks is reduced from 14 to 8, an improvement of 43%.

Note that all three algorithms produce results with densities that are small in comparison with d_{max}, the value of density for a worst possible assignment. Also, for most of the randomly

[1] Modifications were made for some of these to remove incoming/outgoing nets that have only one pin in the channel. This does not affect the results.

Test Problem	density	number of tracks	Iterative Improvement		Simulated Annealing
			Version–D	Version–C	
Deutsch	19	19	14	11	11
YK.3a	14	14	10	8	8
YK.3b	11	11	9	9	9
YK.4b	17	17	15	13	11

Table 5.2

generated problems, $d_{\max}$ is very close to n, and in many cases it is equal to n, the worst possible density, whereas assignments obtained by the algorithms have density much smaller than that. For the specially constructed test problems for which the value of optimal density is known, all three algorithms produced optimal results. It is also interesting to observe that for all test problems the results produced by the simulated annealing algorithm are never inferior to the results produced by the iterative improvement algorithms. In general, this will not always be the case since the simulated annealing algorithm might "get stuck" at a local minimum that is inferior to the result produced by an iterative improvement algorithm. Also, as expected, Version–C of the iterative improvement algorithm outperforms Version–D.

The running times for the iterative improvement algorithms are at the order of a few seconds of CPU time. Version–D takes only 2.7 seconds of CPU time for R7 which contains 50 nets and has a channel length of 124 while Version–C requires about 10.0 seconds. The running time for simulated annealing is at the order of one or two minutes of CPU time. For example, the running time for R7 is about 137.7 seconds.

Finally, Fig.5.11(a) shows an example from [KoDr84] while Fig.5.11(b) shows the result obtained by the two algorithms presented in this chapter. The result of [KoDr84] has density 4 and uses 4 tracks while the solution obtained by the algorithms presented in this chapter has density 2 and uses only 3 tracks.

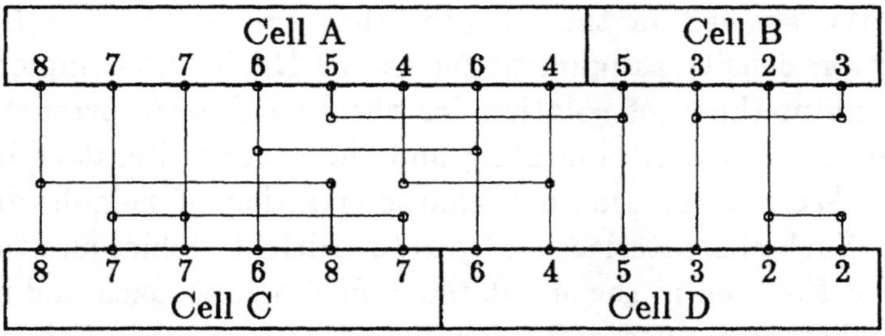

(a)

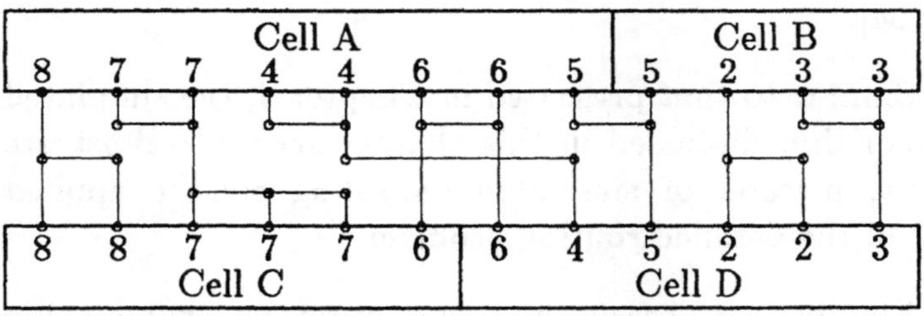

(b)

Fig. 5.11 Comparison of routing results. (a) Routing result obtained by [KoDr84]. (b) Routing obtained by the algorithms presented in this chapter.

5.7. Concluding Remarks

Study of the permutation channel routing problem is motivated by the presence of interchangeable pins in many practical routing situations. For example, logically equivalent pins assigned to different nets can be considered interchangeable pins. Pins on programmable logic devices such as PLA's and ROM's can also be described using the notion of interchangeable pins. The permutation channel routing model gives a natural way to incorporate interchangeable pins into channel routing.

The advantage of permutation channel routing over traditional channel routing is that with the additional flexibility to interchange some of the pins, interconnection can be achieved with fewer tracks. Such savings in the number of tracks can be large, 42% or more in some cases. Thus, the overall chip area can be substantially reduced.

We showed in this chapter that the problem of finding a minimum density assignment for the PCRP is NP-complete. Two different methods of solution for the PCRP were presented. One method is simulated annealing and the other is iterative improvement. We also presented a characterization of neighboring solutions. Such characterization leads to efficient techiniques to update the cost function in the simulated annealing approach and to determine whether a given move is a downhill one or not in the iterative improvement approach. The algorithms for the PCRP produce very good results. In particular, they outperform the algorithm in [KoDr84].

Similar to that presented in Chapter 4, the simulated annealing algorithm discussed in this chapter serves to illustrate further how the method of simulated annealing can be applied to one aspect of the detailed routing problem.

Finally, we observe in this chapter an example illustrating the notion of postponing some of the decisions on circuit/pin placement in placement–routing problems so that additional flexibilites can be exploited in the routing stage in order to obtain better overall solutions. Specifically, such an approach is taken for the channel routing problem by generalizing the problem to a permutation channel routing problem. This general approach can also be applied to other circuit layout problems, such as switchbox routing, river routing, and Gate Matrix layout.

CHAPTER 6

PLA FOLDING

6.1. Introduction

Programmable Logic Arrays (PLA's) are structured logic arrays which implement multiple output combinational logic functions. They are perhaps the most popular circuit modules that implement two–level logic functions. In general, a PLA can be described in symbolic form by a matrix called ***personality matrix***. Most personality matrices are very sparse. Consequently, a direct implementation of a PLA will result in a rather significant waste of chip area. Such waste will reduce circuit yield and degrade time performance of the PLA. PLA folding [HaNS82, DeSa83, LeLi84, WoLL87] is a general technique to reduce the total area of a PLA. By a *simple folding* of the columns of a PLA, we mean to let two input variables or two output functions share one column. In a *multiple folding* of the columns of a PLA, we allow two or more input variables or two or more output functions to share one column. It is customary to disallow an input variable and an output function to share the same column. For the PLA in Fig.6.1(a), Fig.6.1(b) shows a simple folding of the columns and Fig.6.1(c) shows a multiple folding of the columns of the PLA. The objective of PLA folding is to find a simple/multiple folding of the columns of a PLA such that the total number of columns in the folded PLA is minimized.

The problem of obtaining an optimal folding has been shown to be NP–complete [LuVV82]. Consequently, most research efforts have been restricted to heuristic algorithms [HaNS82, DeSa83] which do not carry out a thorough search of the solution space, and to branch–and–bound algorithms [Gras82 , LeLi84] which can only handle PLA's of moderate size (no more than 50 to 100 input/output lines). Furthermore, most of these efforts have been concentrated on simple folding. Besides [Gree76, DeSa83,

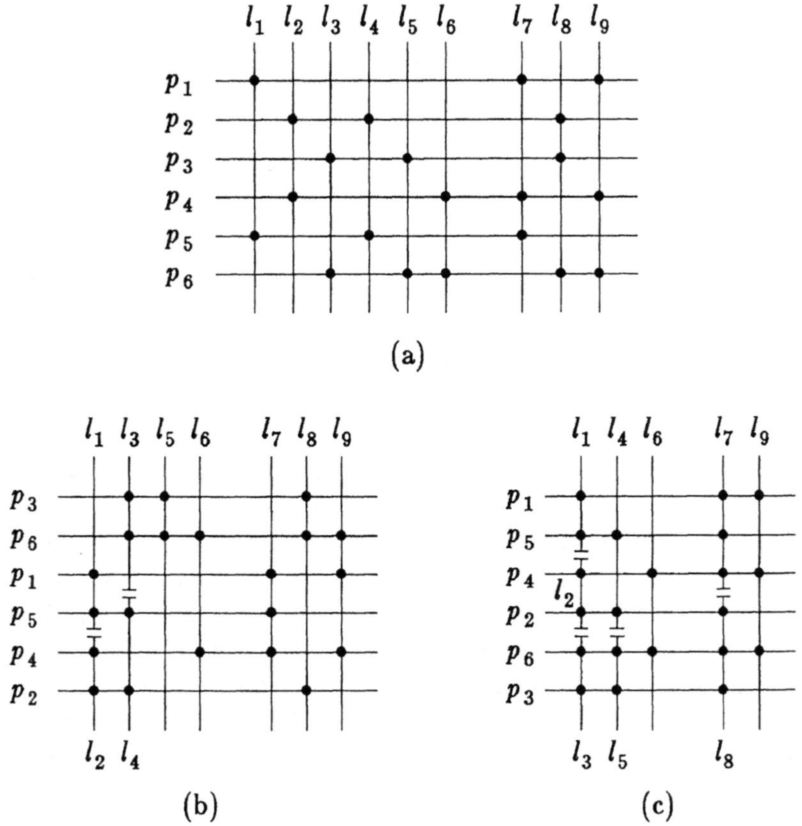

Fig. 6.1 PLA folding. (a) A PLA. (b) Simple folding of the columns of the PLA in (a). (c) Multiple folding of the columns of the PLA in (a).

YuWi85], relatively little has been done for multiple folding.

In this chapter, we present a simulated annealing PLA folding algorithm [WoLL86, WoLL87] which can perform multiple folding as well as simple folding. As a matter of fact, in this algorithm, the computational effort for obtaining a multiple folding is a little bit less than that for obtaining a simple folding because in the problem formulation, simple folding is considered to be a special case of multiple folding with additional constraints. We also show how to extend the algorithm to handle multiple constrained folding.

In applying the method of simulated annealing to the solution of combinatorial optimization problems, there are two major considerations, namely, formulation of the solution space and definition of the notion of neighborhood in the solution space. In most of the work on PLA folding, algorithms examine partial solutions and try to expand a partial solution in a step–by–step fashion. Such an approach is not suitable for the simulated annealing setting because a simulated annealing algorithm needs to examine *complete* solutions one after another. In [WoLL86] and [WoLL87], the column folding problem is formulated as a row permutation problem which leads to a simple and efficient way in obtaining simple/multiple foldings of PLA's.

The simulated annealing folding algoritm compares well with other algorithms. In the case of multiple folding, the algorithm was compared with a branch and bound optimal algorithm due to Skrien [Skri85] and an implementation of the unconstrained folding algorithm due to De Micheli and Sangiovanni [DeSa83]. In the case of simple folding, the algorithm was compared with the a branch and bound optimal algorithm due to Lewandowski and Liu [LeLi84]. The simulated annealing folding algorithm obtained optimal results in most of the test problems and also consistently outperformed De Micheli and Sangiovanni's algorithm.

This chapter is organized as follows. A precise formulation of the PLA folding problem is given in Section 6.2. Section 6.3 describes in detail the simulated annealing algorithm for multiple PLA folding. Sections 6.4 and 6.5 describe how the algorithm can be extended to handle realization of multiple–folded PLA and various constraints on the folding problem. Specialization of the algorithm to the problem of simple folding is described in Section 6.6. Experimental results are presented in Section 6.7. Finally, some concluding remarks are given in Section 6.8.

6.2. The PLA Folding Problem

A *personality matrix* is a symbolic representation of a PLA which has one column for each input variable and output function, and one row for each product term. The *ij th* entry of the matrix is a 1 if the *j th* input variable (or its complement) or the *j th* output function is present in the *i th* product term. Otherwise, the

$$M = \begin{array}{c} \begin{array}{ccccccccc} l_1 & l_2 & l_3 & l_4 & l_5 & l_6 & l_7 & l_8 & l_9 \end{array} \\ \left[\begin{array}{cccccc|ccc} 1 & 0 & 0 & 0 & 0 & 0 & 1 & 0 & 1 \\ 0 & 1 & 0 & 1 & 0 & 0 & 0 & 1 & 0 \\ 0 & 0 & 1 & 0 & 1 & 0 & 0 & 1 & 0 \\ 0 & 1 & 0 & 0 & 0 & 1 & 1 & 0 & 1 \\ 1 & 0 & 0 & 1 & 0 & 0 & 1 & 0 & 0 \\ 0 & 0 & 1 & 0 & 1 & 1 & 0 & 1 & 1 \end{array} \right] \begin{array}{c} p_1 \\ p_2 \\ p_3 \\ p_4 \\ p_5 \\ p_6 \end{array} \end{array}$$

Fig. 6.2 Personality matrix M of the PLA in Fig. 6.1(a).

ij *th* entry is a 0. The personality matrix of the PLA in Fig.6.1(a) is shown in Fig.6.2. Let M be the personality matrix of a PLA. Let n be the number of product terms, m_1 be the number of input variables, and m_2 be the number of output functions. Let $m = m_1 + m_2$. Clearly, M is an $n \times m$ matrix. An input column will be referred to as an *input line* and an output column will be referred to as an *output line*. A *line* is either an input line or an output line. Let l_i denote the i *th* column of M, $1 \leq i \leq m$. Throughout our discussion, we shall have $l_1, l_2, ..., l_{m_1}$ as the input lines, and $l_{m_1+1}, l_{m_1+2}, \cdots, l_m$ as the output lines. Let $\{p_1, p_2, \cdots, p_n\}$ be the set of product terms of the given PLA. We shall assume that row i of M corresponds to product term p_i, $1 \leq i \leq n$.

Given a permutation π of the rows of M, consider the matrix M_π obtained by permuting the rows of M according to π. Clearly, M_π corresponds to a different realization of the PLA. Fig.6.3(a) shows the matrix M_π for the personality matrix M in Fig.6.2 with $\pi = 1\ 5\ 4\ 2\ 6\ 3$. Fig.6.3(b) shows a different realization of the PLA in Fig.6.1(a) corresponding to M_π. With respect to a permutation π, each line l_i can be represented by an interval $I_i = [a_i, b_i]$ where a_i and b_i are the positions of the topmost and the bottommost 1 in column i of M_π, respectively. I_i is an *input interval* if l_i is an input line and is an *output interval* if l_i is an output line. Fig.6.4 shows the input/output intervals for the example in Fig.6.3. (The heavy line segments in Fig.6.4 correspond to the input/output intervals.) Two input (output) lines l_i and l_j are said to be *foldable* if I_i and I_j

$$
\begin{array}{c}
 \\
M_\pi = \\
\end{array}
\begin{array}{c}
\begin{array}{cccccccccc} l_1 & l_2 & l_3 & l_4 & l_5 & l_6 & l_7 & l_8 & l_9 & \end{array} \\
\left[\begin{array}{ccccccccc}
1 & 0 & 0 & 0 & 0 & 0 & 1 & 0 & 1 \\
1 & 0 & 0 & 1 & 0 & 0 & 1 & 0 & 0 \\
0 & 1 & 0 & 0 & 0 & 1 & 1 & 0 & 1 \\
0 & 1 & 0 & 1 & 0 & 0 & 0 & 1 & 0 \\
0 & 0 & 1 & 0 & 1 & 1 & 0 & 1 & 1 \\
0 & 0 & 1 & 0 & 1 & 0 & 0 & 1 & 0
\end{array}\right]
\end{array}
\begin{array}{c}
p_1 \\ p_5 \\ p_4 \\ p_2 \\ p_6 \\ p_3
\end{array}
$$

(a)

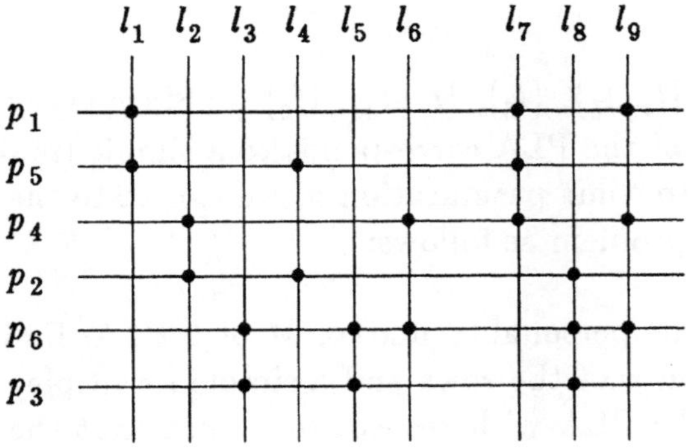

(b)

Fig. 6.3 M_π and its PLA realization. (a) M_π obtained by permuting the rows of the matrix M in Fig. 6.2. (b) A different realization of the PLA in Fig. 6.1(a) corresponding to M_π.

are disjoint. Clearly, with respect to π, two foldable input (output) lines can be assigned to share the same column. For example, $\{l_1, l_3\}$, $\{l_7, l_8\}$ are two pairs of foldable lines, but neither $\{l_2, l_6\}$ nor $\{l_5, l_7\}$ are foldable. A ***multiple folding with respect to*** π is a partition $F_\pi = \{C_1, C_2, \ldots, C_k\}$ of the lines such that each C_i either consists of a single line or is a set of pairwise foldable lines. A partition F_π is a ***simple folding with respect to*** π if each C_i consists of at most two lines; *i.e.* $|C_i| \leq 2$. In both cases, each C_i corresponds to a column in the folded PLA. The *size* of the simple/multiple folding is defined as the number of blocks in the partition F_π, $|F_\pi|$, which is equal to the number of columns in the folded PLA. Fig.6.5 shows a folded version of the example in Fig.6.4 with $F_\pi =$

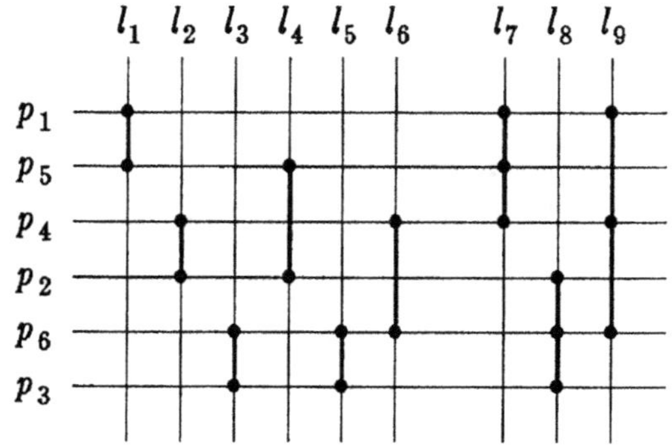

Fig. 6.4 Input/Output intervals.

$\{ \{l_1, l_2, l_3\}, \{l_4, l_5\}, \{l_6\}, \{l_7, l_8\}, \{l_9\} \}$. Since every simple (multiple) folding of the PLA corresponds to a simple (multiple) folding with respect to some permutation π, we can state the simple (multiple) folding problem as follows:

> Given the personality matrix M of a PLA, find a permutation π of the rows and a simple (multiple) folding F_π of the PLA with respect to π such that the size of F_π is minimum.

6.3. The PLA Folding Algorithm

In this section, we present a simulated–annealing PLA folding algorithm. This algorithm can be applied to both simple and multiple folding. To simplify our presentation, we only describe the algorithm in terms of multiple folding in this section, and shall discuss the case of simple folding in Section 6.6. We now give a detail description of the folding algorithm.

6.3.1. Solution Space

In general, for a given permutation π of the rows of M, there are many different foldings with respect to π that are of different sizes. We shall show later in this section that there is an efficient algorithm to find a minimum size folding (denoted $F_\pi^{\min}$) among these foldings. Clearly, to solve the PLA folding problem, it is

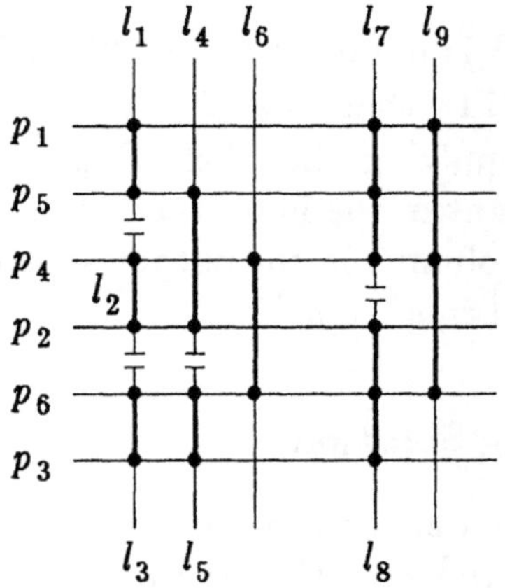

Fig. 6.5 A folded version of the example in Fig. 6.4.

sufficient to consider only the set of all $F_\pi^{\min}$ for all permutations π of the rows of M. The fact that we can easily obtain $F_\pi^{\min}$ from π allows us to represent folding solutions by permutations of the rows of M. Given a permutation π, it is clear that the problem of finding $F_\pi^{\min}$ is equivalent to the problem of separately packing the input intervals and the output intervals into as few columns as possible. Fortunately, this packing problem can be solved using an algorithm in [HaSt71]. In fact, we can view the input intervals as nets in a channel routing problem and the output intervals as nets in another channel routing problem where there are no vertical constraints [YoKu82] between the nets in both problems. The "Left Edge Algorithm" in [HaSt71] can then be applied separately to the input intervals and the output intervals to produce an optimal packing of the input intervals and an optimal packing of the output intervals. These two packings together yield $F_\pi^{\min}$.

We say row i of M_π *intersects* interval I_j if $i \in I_j$. Let the *input density* of row i of M_π (denoted d_i^I) be the number of input intervals it intersects, and let the *input density* of M_π (denoted d^I) be the maximum input density of all the rows. Similarly, we can define the *output density* of row i of M_π (denoted d_i^O) and the *output density* of M_π (denoted d^O) in terms of the output intervals. For the example in Fig.6.6, the numbers on the left are the d_i^I's and the numbers on the right are the d_i^O's. In this example, d^I is 3 and

d^O is 2. Since $F_\pi^{\min}$ is obtained by the "Left Edge Algorithm," it follows from [HaSt71] that the size of $F_\pi^{\min}$ can be obtained without explicitly applying the "Left Edge Algorithm." In fact, the number of columns in the input section of the folded PLA is d^I and the number of columns in the output section of the folded PLA is d^O. Hence $|F_\pi^{\min}| = d^I + d^O$.

6.3.2. Neighboring Solutions

A permutation π of the rows of M is a one to one function $\pi : \{1,2,...,n\} \rightarrow \{1,2,...,n\}$ such that the i *th* row of M_π is the $\pi(i)$ *th* row of M. We shall also use the notation $\pi(1)\pi(2) \cdots \pi(n)$ to denote a permutation π. Given a permutation π of the rows of M, a *move* is an interchange of two function values $\pi(i)$ and $\pi(j)$ for some i,j. Clearly, it is possible to go from any permutation to any other via a sequence of moves. Two permutations are said to be *neighbors* if one can be obtained from the other by such a move. Two folding solutions are said to be *neighbors* if their corresponding permutations are neighbors. Clearly, if π_1 and π_2 are neighbors, we can obtain either one of the matrices M_{π_1} and M_{π_2} from the other by interchanging two of the rows.

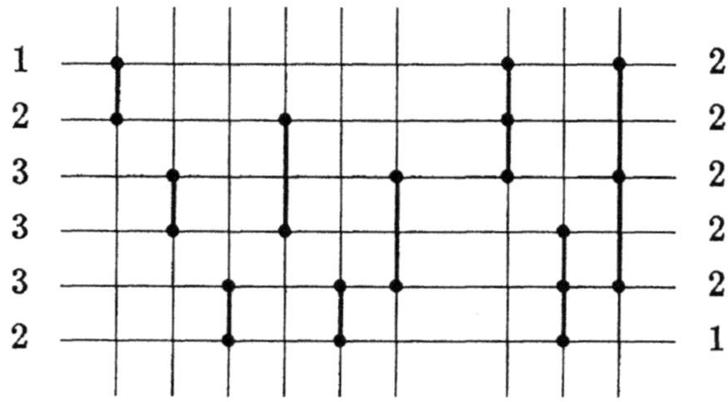

Fig. 6.6 Input/output densities of rows.

6.3.3. Cost Function

As mentioned in Section 6.2, the size of $F_\pi^{\min}$ can be computed directly from the formula $|F_\pi^{\min}| = d^I + d^O = \max_i \{d_i^I\} + \max_i \{d_i^O\}$ without explicitly applying the "Left Edge Algorithm." This allows us to reduce the computation time substantially. For any permutation π, instead of using the objective function $d^I + d^O$ as the cost function, we use the following cost function which is a more refined measure of the quality of intermediate solutions:

$$C = (d^I + d^O)^2 + \frac{\lambda}{n} \sum_{i=1}^{n} (d_i^I + d_i^O)^2 = D^2 + \lambda E$$

where $D = d^I + d^O$, $D_i = d_i^I + d_i^O$, $E = \frac{1}{n} \sum_{i=1}^{n} (d_i^I + d_i^O)^2$, and λ is a constant. The first term in the cost function is the square of D which is the objective function to be minimized. To see why we use D^2 instead of D in the cost function, suppose that a move increases the value of D by t, the corresponding increase, Δ_D, would be $2tD + t^2$. This means that we are placing a higher penalty on solutions with large D and lower penalty on solutions with small D. The second term in the cost function is λE. Note that E and D^2 are, respectively, the average value and the maximum value of all the D_i^2 's. In general, we expect that among the solutions with the same value D^2, those with smaller values of E are "closer" to solutions with smaller values of D^2, and hence smaller values of D. The value of λ is used to control the relative importance of D^2 and E. Since D is the quantity we want to minimize, it will assume greater importance *i.e* $\lambda \leq 1$. Setting $\lambda = 1$ means assigning equal importance to D^2 and E, and setting $\lambda = 0$ reduces the cost function to D^2 which, in general, is not sufficient to measure the quality of intermediate solutions. Experimental results indicate that $\lambda = 0.5$ is a good choice.

Note that the moves only interchange two rows at a time. In general, only some of the intervals I_j's, and hence only the values of some of the D_i's, are changed. Therefore, instead of recomputing all the D_i's after every move, we can update only those that are changed and thus further reduce computation time.

6.3.4. Annealing Schedule

The temperature schedule is of the form $T_k = r \cdot T_{k-1}$, for $k = 1, 2, 3, \ldots$. Experimental results indicate that setting r to 0.8 produces very satisfactory results. Since each move is a local perturbation of a solution, we expect that in general, Δ_D, the change in D cannot be too big. To choose a suitable value of T_0, we assume that $\Delta_D \leq K$ for some constant K. For a move that increases D by K, the corresponding change in the cost function is $\Delta \simeq 2KD + K^2$ and the probability of accepting this move is $e^{-\Delta/T}$. Thus for there to be a reasonable probability of acceptance at high temperatures, we should have $T_0 = O(D)$. In fact, T_0 is chosen to be of the form $\lambda_T \cdot m$. Recall that m is the total number of input/output lines in the PLA and is clearly an upper bound of D.

The folding algorithm can start with any initial permutation. The default initial permutation is the identity permutation $\pi = 1 2 3 \cdots n$. Hence, initially we have $M_\pi = M$. At each temperature, we try enough moves until either there are N downhill moves or the number of moves exceeds $2N$ where $N = \lambda_N N_1$ with $\lambda_N < 1$ and $N_1 = n(n-1)/2$ is the number of neighbors. We terminate the annealing process if the number of accepted moves is less than 5% of all the moves made at the temperature or the temperature is low enough.

6.4. Multiple–Folded PLA Realization

In any physical realization of a folded PLA, we must ensure that the input/output lines can be connected to outside circuitry. In the case of simple folding, the realization is relatively straightforward because at most two lines are folded together and connection to outside circuitry can be done from the top and the bottom of the array. The realization of multiple–folded PLA is more complicated. Two possible architectures were considered in [DeSa83]. They will be referred to as A_1 and A_2. In these architectures, horizontal connection lines are introduced to route the input and output lines to outside circuitry. The connection lines for the input lines all leave the PLA on one side while those for the output lines all leave on the other side. In the first architecture A_1, there is a third layer referred to as the interconnection layer. All the connection lines will run above the product term rows of the folded PLA using this layer. In the second architecture A_2, the connection lines run parallel and alternately to the product term rows of the folded PLA. See Fig.6.7 for an illustration of the connection lines in these two architectures. Both architectures require the total number of input lines and the total number of output lines be at most equal to the number of product terms. Most PLA's satisfy this assumption.

We define the *position* of a connection line for an input/output line to be the index i if the connection line runs above row i of the folded PLA in the case of architecture A_1, or runs between row i and row $i+1$ of the folded PLA in the case of architecture A_2. A *connection-line assignment* is a specification of the positions of the connection lines for all the input/output lines. We define an *input connection assignment* as a one-to-one function $f : \{1,2,...,m_1\} \rightarrow \{1,2,...,n\}$ and an *output connection assignment* as a one-to-one function $g : \{1,2,...,m_2\} \rightarrow \{1,2,...,n\}$. $f(i)$ specifies the position of the connection line for the input line l_i and $g(i)$ specifies the position of the connection line for the output line l_{m_1+i}. It is easy to modify the simulated annealing algorithm to perform multiple folding together with connection-line assignment. Given a permutation π and a pair of input/output connection

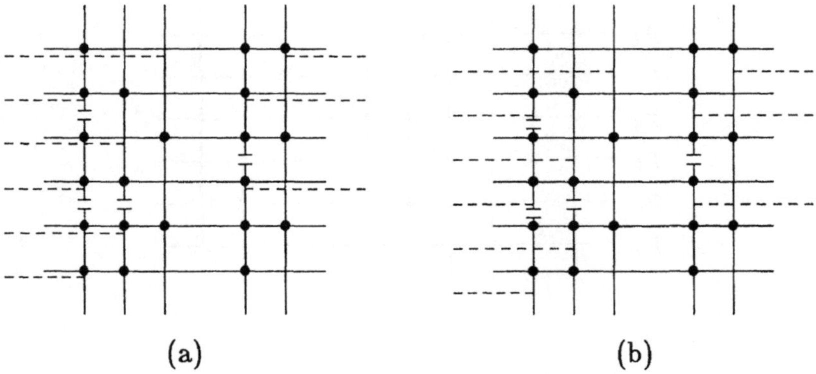

Fig. 6.7 Realization of multiple-folded PLA. (a) Architecture A_1. (b) Architecture A_2. Dashed lines are connection lines.

assignment functions f and g, we have to slightly modify the definition of the intervals I_i's for each line l_i. The definitions are also slightly different for the two architectures. For a line l_i, let a_i and b_i be defined as before as the topmost and bottommost 1 in column i of M_π. If l_i is an input line, we let c_i be $f(i)$ in the case of architecture A_1 and be $f(i) + 0.5$ in the case of architecture A_2. If l_i is an output line, we let c_i be $g(i-m_1)$ in the case of architecture A_1 and be $g(i-m_1) + 0.5$ in the case of architecture A_2. I_i is then defined as the interval $[\ \min\{a_i, b_i, c_i\}, \max\{a_i, b_i, c_i\}\ ]$. Fig.6.8 illustrates the definition of input/output intervals for architecture A_1. For the example in Fig.6.8(b) with the functions π, f and g given as in Fig.6.8(a), the values of the $f(i)$'s are represented

i	1	2	3	4	5	6
$\pi(i)$	1	5	4	2	6	3
$f(i)$	2	4	6	1	3	5
$g(i)$	2	5	6	–	–	–

(a)

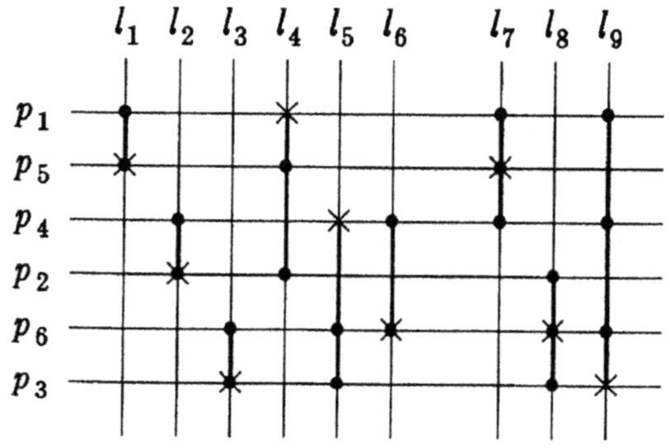

(b)

Fig. 6.8 Modified input/output intervals with respect to $\pi_1 f_1$ and g.

by the $\times$'s in the first 6 columns and the values of the $g(i)$'s are represented by the $\times$'s in the last 3 columns. The input/output intervals are those heavy line segments shown in the figure.

As before, for a given permutation π and a pair of input/output connection assignment functions f and g, we can apply the "Left Edge Algorithm" to the set of intervals I_i's to obtain a minimum size folding, and the size of the folding is again the sum of the two densities d^I and d^O. Hence folding solutions (with connection–line assignment) are represented by the set of all triplets of functions (π,f,g) where π, f and g are one–to–one functions from $\{1,2,...,n\}$, $\{1,2,...,m_1\}$ and $\{1,2,...,m_2\}$ to $\{1,2,...,n\}$, respectively. Fig.6.9 shows a folded version of the example in Fig.6.8 using architecture A_1. This is a minimum size folding because d^I is 4, d^O is 2 and the number of columns in the folded PLA is 6, which is $d^I + d^O$.

In addition to interchanging the function values of π, we extend the definition of a move to include the interchange of the function values of f and g, and refer to this type of moves as M1. We also introduce M2, a new type of move, to reflect local changes of the functions f and g. M2 is only needed when $m_1 < n$ or $m_2 < n$. M1 and M2 are defined as follows:

M1: Interchange two function values of π or f or g.
M2: Change one function value of f or g such that the function remains a one–to–one function.

6.5. Constrained Multiple Folding

A PLA is only a part of the entire circuit. It needs to be connected to the the outside circuitry which in general imposes some constraints on the positions of the connection lines or the positions of the input/output lines. The folding algorithm presented here is flexible enough to handle different kinds of constraints such as those considered in [DeSa83] and [YuWi85]. Most of these constraints can be expressed in terms of constraints on the functions π, f, or g as demonstrated in the following discussion.

(1) *Bounded Product–Term Positions:* Given a folded solution, each row of the folded PLA corresponds to a product term.

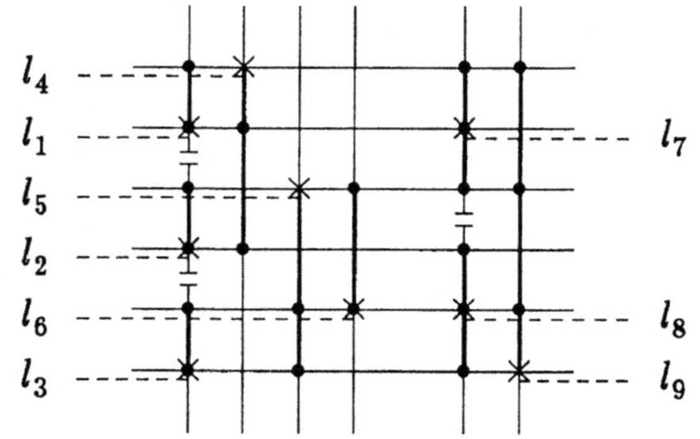

Fig. 6.9 A folded version of the eaxmple in Fig.6.8 – Architecture A_1.

The *position* of a product term p_i is defined to be the index j if row j of the folded PLA corresponds to p_i. Constraints on the positions of the product terms defined by two functions which are the lower/upper bound functions L/U from $\{1,2,...,n\}$ to $\{1,2,...,n\}$. It is required that the position of the product term p_i must lies between $L(i)$ and $U(i)$, $1 \leq i \leq n$. Clearly, such constraints can be expressed in terms of constraints on π as follows: $L(i) \leq \pi^{-1}(i) \leq U(i)$, $1 \leq i \leq n$.

(2) *Bounded Connection–Line Positions:* Constraints on the positions of the connection lines are defined by a pair of lower/upper bound functions L/U from $\{1,2,...,m\}$ to $\{1,2,...,n\}$. For all i, the position of the connection lines for the line l_i must lie between $L(i)$ and $U(i)$. It is clear that these constraints can be expressed in terms of the functions f and g as follows: $L(i) \leq f(i) \leq U(i)$, $1 \leq i \leq m_1$, and $L(i) \leq g(i-m_1) \leq U(i)$, $m_1+1 \leq i \leq m$.

(3) *Ordered Connection–Line Positions:* Constraints on the relative ordering of the positions of the connection lines are defined by two partially ordered set P_1 and P_2. P_1 has m_1 vertices corresponding to the m_1 input connection lines and P_2 has m_2 vertices corresponding to the m_2 output connection lines. Such constraints can be expressed in terms of f and g as follows: $f(x) < f(y)$ if $x < y$ in P_1, and $g(x) < g(y)$ if $x < y$ in P_2.

(4) *Top–Bottom Input/Output Lines Restrictions:* These constraints require that some specific input/output lines must appear at the top and some others must appear at the bottom of the folded PLA. We can handle these constraints by adding two fictitious rows, a row 0 and a row $n+1$, to M. The i *th* entry in row 0 (row $n+1$) is 1 if the input/output line i must appear at the top (or bottom), and is 0 otherwise. We shall only consider permutations π that do not permute these two special rows.

We note that there are two alternative approaches to modify the basic folding algorithm to perform constrained folding. For a given set of constraints, a triplet of functions (π, f, g) is said to be a *valid solution* if these functions satisfy the given constraints, and is said to be an *invalid solution* otherwise. The first approach is to reduce the solution space to the set of all valid solutions and only consider those moves that move from valid solutions to valid solutions. Another approach is to use the original solution space and add a penalty term to the cost function so that invalid solutions will be discouraged. The magnitude of the penalty term depends on how severe the violation of the constraints is. The constrained folding problem with top–bottom input/output lines restrictions can be best handled by the first approach. We now describe the penalty terms for the other three types of constrained folding if the second approach is used.

(1) For the constrained folding problem with bounded product–term positions, the penalty term in the cost function is proportional to

$$\sum_{i=1}^{n} \Psi(\pi^{-1}(i), J_i)$$

where J_i is the interval $[L(i), U(i)]$, and the function Ψ is defined as follows: $\forall\, x,\ a,\ b$ such that $a < b$, $\Psi(x,[a,b]) = 0$ if $x \in [a,b]$, and $\Psi(x,[a,b]) = \min\{|x-a|, |x-b|\}$ otherwise.

(2) For the constrained folding problem with bounded connection–line positions, the penalty term in the cost function is proportional to

$$\sum_{i=1}^{m_1} \Psi(f(i), J_i) + \sum_{i=m_1+1}^{m} \Psi(g(i-m_1), J_i)$$

where $J_i = [L(i), U(i)]$.

(3) For the constrained folding problem with ordered connection–line positions, the penalty term in the cost function is proportional to

$$\sum_{(x,y)\in S_1} (f(x) - f(y)) + \sum_{(x,y)\in S_2} (g(x) - g(y))$$

where $S_1 = \{(x,y) \mid x,y \in P_1,\ x < y,\ \text{and } f(x) > f(y)\}$, and $S_2 = \{(x,y) \mid x,y \in P_2,\ x < y,\ \text{and } g(x) > g(y)\}$.

6.6. Simple Folding

We now describe how to apply the simulated annealing algorithm to the problem of simple folding of PLA's. The only difference between simple folding and multiple folding is the additional restriction that each column in the folded PLA contains at most two lines. We shall show that the solution space for simple folding is the same as that for multiple folding. We can also use the same definition of neighboring solutions and the same basic annealing schedule. The only difference is in the cost function.

As in the case of multiple folding, for a given permutation π of the rows of M, there are many different simple foldings of different sizes. We can use the set of all permutations of the rows of M as the solution space provided that we have an efficient algorithm to obtain a minimum size simple folding with respect to a given permutation π. Let G_π be an undirected graph where the vertices correspond to the lines and there is an edge between vertex i and vertex j if and only if the two lines l_i and l_j are foldable. A matching in G_π defines a set of pairs of lines that can be folded together, and hence a simple folding. Let k be the cardinality of the matching. The number of unmatched vertices is then equal to $m-2k$. Hence the size of the simple folding is equal to $k + m-2k$, which can be simplified to $m-k$. Consequently, a maximum cardinality matching in G_π defines a minimum size simple folding of the PLA with respect to π. For the example in Fig.6.10(a), Fig.6.10(b) shows the graph G_π and a maximum cardinality matching in G_π, and Fig.6.10(c) shows a minimum size folding corresponding to the matching. There are efficient algorithms to find a maximum cardinality matching in a graph [MiVa80]. Hence, for every permutation

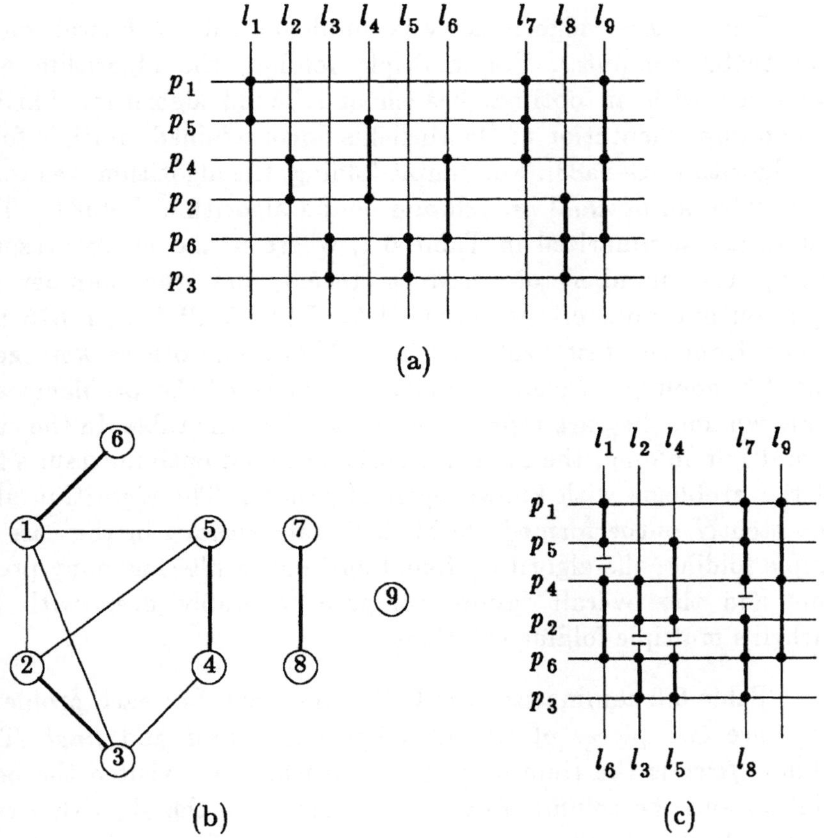

Fig. 6.10 Maximum cardinality matching and simple folding. (a) The PLA in Fig. 6.1(a) with respect to $\pi = 1\ 5\ 4\ 2\ 6\ 3$. (b) G_π and a maximum cardinality matching. (c) A simple folding corresponding to the matching in (b).

π, we can efficiently obtain a minimum size folding with respect to π. Therefore, as in the case multiple folding, we can use the set of all permutations of the rows of M as the solution space. For a given permutation π, let k_π be the number of edges in a maximum cardinality matching in G_π. Then $m-k_\pi$ is the size of a minimum size folding with respect to π. Thus, we use the cost function $C = (m-k_\pi)^2$.

6.7. Experimental Results and Discussions

The folding algorithm was implemented in Pascal on a PYRAMID computer. For multiple folding, the algorithm was compared with an optimal branch and bound algorithm [Skri85] and an implementation of De Micheli's unconstrained multiple folding algorithm [DeSa83]. For simple folding, the algorithm was compared with an optimal branch and bound algorithm [LeLi84]. The results are summarized in Table 6.1, where n and m are, respectively, the number of product terms and the number of input/output lines of the given PLA. The 15 PLA's p1–p15 are chosen from the test problems in [LeLi84]. The others were generated randomly. Optimal results for some of the problems are unknown and they are represented by a "–" in the table. In the case of multiple folding, the folding algorithm found optimal results for all the problems with known optimal results. The algorithm also consistently outperformed De Micheli's algorithm. In the case of simple folding, the algorithm found optimal results for many problems and the overall results compare favorably even with De Micheli's multiple folding solutions.

Table 6.2 summarizes the CPU time used. For each problem, there are two pieces of timing information: *first* and *final*. The column *first* is the time when the algorithm first visited the best solution and the column *final* is the time when the algorithm terminated (after about 10 temperature changes). The results indicate that the algorithm is capable of visiting very good solutions at a very early stage. Therefore, we can expect to have good results even when an annealing schedule with less number of iterations (and hence with less CPU time) is used. The results also indicate that the algorithm is more efficient for multiple folding than for simple folding. This is due to the fact that the cost function computation in the case of multiple folding is very efficient, the algorithm only have to update the local densities after each move. Whereas in the case of simple folding, the procedure for maximum cardinality matching is used twice for every cost function computation. A possible way to speed up the algorithm for simple folding is to design a fast and good heuristic for maximum cardinality matching.

PLA (nxm)	Number of Columns in the Folded PLA				
	Simple Folding		Multiple Folding		
	SA	OPT	SA	OPT	DS
p1 (14x15)	11	11	10	10	11
p2 (17x14)	12	12	12	12	12
p3 (6x10)	5	5	5	5	5
p4 (8x14)	8	8	6	6	6
p5 (21x18)	13	13	13	13	14
p6 (27x30)	21	21	18	18	20
p7 (22x33)	29	29	29	29	30
p8 (24x40)	37	37	37	37	38
p9 (20x32)	26	25	24	24	26
p10 (15x38)	34	34	34	34	34
p11 (46x40)	32	31	32	–	32
p12 (70x50)	35	35	32	–	35
p13 (59x50)	38	36	32	–	35
p14 (33x42)	22	22	14	–	17
p15 (19x23)	13	13	10	10	10
p16 (20x80)	62	–	62	–	66
p17 (20x80)	45	–	43	–	52
p18 (20x100)	51	–	48	–	63
p19 (20x120)	60	60	47	–	76
p20 (20x120)	60	60	53	–	72
p21 (17x68)	34	–	19	19	37
p22 (20x76)	38	–	27	–	44

Table 6.1

PLA	CPU Time (in sec.)			
	Simple Folding		Multiple Folding	
	first	final	first	final
p1	0.7	12.0	0.5	3.5
p2	0.6	14.0	0.4	5.0
p3	0.2	1.6	0.1	0.7
p4	0.1	1.3	0.1	1.0
p5	3.0	35.0	3.0	9.0
p6	21.0	184.0	12.0	20.0
p7	7.0	109.0	1.0	15.0
p8	2.0	170.0	0.7	22.0
p9	10.0	83.0	5.0	11.0
p10	4.0	66.0	0.4	7.0
p11	161.0	1004.0	51.0	77.0
p12	1073.0	1503.0	207.0	222.0
p13	2.0	1500.0	149.0	161.0
p14	126.0	443.0	16.0	28.0
p15	4.0	46.0	0.8	6.0
p16	30.0	617.0	22.0	24.0
p17	512.0	649.0	5.0	16.0
p18	401.0	1212.0	4.0	17.0
p19	82.0	1497.0	15.0	18.0
p20	139.0	1502.0	15.0	19.0
p21	1.6	204.0	19.0	20.0
p22	18.0	621.0	12.0	13.0

Table 6.2

6.8. Concluding Remarks

Programmable Logic Arrays in general are very sparse. Direct implementation of a PLA will result in a rather significant waste of chip area. PLA folding is a general technique to reduce the total area of a PLA. In this chapter, we presented a simulated annealing PLA folding algorithm for simple as well as multiple column folding. The column folding problem was formulated as a row permutation problem which lead to a simple and efficient way of obtaining simple/multiple foldings of PLA's. We also showed how to extend the algorithm to handle multiple constrained folding. Experimental results indicate that the algorithm compares well with other algorithms [DeSa83, LeLi84, Skri85]. The algorithm obtained optimal results in most of the test problems and also consistently outperformed the algorithm reported in [DeSa83].

CHAPTER 7

GATE MATRIX LAYOUT

7.1. Introduction

In this chapter, we consider the layout of MOS circuits in the style of Gate Matrix. This layout style was introduced by Lopez and Law [LoLa80] in 1980 for the layout of custom CMOS circuits. It is very similar to other one–dimensional layout styles, such as Weinberger arrays, that were studied in [Wein67, OhMK79, KaFu79]. Since it uses a very simple and regular structure, this layout style greatly simplifies the layout procedure. It is also adaptable to a team effort and can accommodate changes in the design or design rules. Various algorithms for the Gate Matrix layout problem have been developed [LoLa80, Wing82, Wing83, Li83, WiHW85]. These algorithms are based on an earlier algorithm by Ohtsuki *et al* [OhMK79] for the One–Dimensional Logic Assignment Problem.

A general formulation of the Gate Matrix layout problem can be described as follows [WiHW85]: We are given a *circuit schematic*, an example of which is shown in Fig.7.1(a). We want to obtain a layout of the given circuit in the style of Gate Matrix. Fig.7.1(c) and Fig.7.1(e) show two possible layouts of the circuit in Fig.7.1(a). The goal is of course to minimize the total area of the layout.

We now describe the Gate Matrix layout style. The columns are equally spaced polysilicon lines which serve the dual role of transistor gates and interconnections. For simplicity, we shall refer to these as the *gate lines*. For example, in Fig.7.1(b), the gate line labelled c serves the dual role of being a transistor gate for net 2 and an output terminal for net 1. The rows are diffusion or metal lines. Transistors are placed at the intersections of rows and columns and these are formed by placing a horizontal strip of

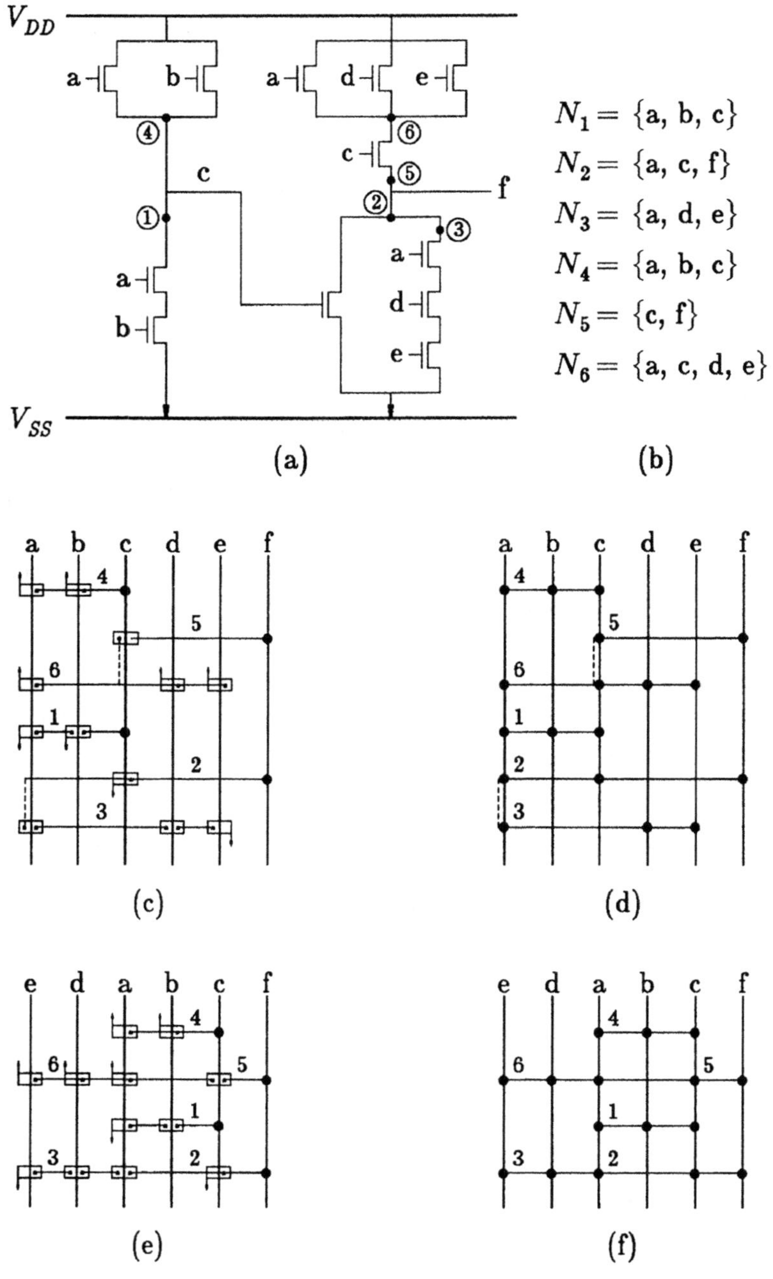

Fig. 7.1 Gate Matrix layout of a CMOS circuit. (a) Circuit schematic. (b) Net–list. (c) A possible layout. (d) Abstract layout. (e) A better layout. (f) Abstract layout.

diffusion at the corresponding intersection points. Horizontal metal line segments are used for interconnection between transistors. Similar to [WiHW85], we shall assume that power and ground connections are placed on a second layer of metal. The routing problem for these connections will not be discussed here.

The Gate Matrix layout style can be applied to any MOS circuit in general. The given circuit schematic can either be a network of logic gates or, more generally, be in the form of a transistor network such as that in Fig.7.1(a). In the layout of a CMOS circuit such as that shown in Fig.7.1(a), we shall assume that there is a power line V_{DD} running across the top of the matrix and a ground line V_{SS} across the bottom. All the p–transistors are placed in the top half of the matrix and all the n–transistors in the lower half.

In this chapter, we present a simulated annealing approach to the Gate Matrix layout problem [Leon86b]. We first present a generalization of the formulation of [WiHW85] so as to obtain an integrated approach to the problem. We then show that the generalized formulation is indeed superior to that of [WiHW85]. Finally, we present a simulated annealing layout algorithm that takes into account practical constraints on the layout problem, and solves the global optimization problem in one integrated step. The advantage of the layout algorithm is that it performs the following optimization steps simultaneously: (1) minimizing the number of tracks, (2) allowing "dynamic binding" of transistors to nets, and (3) considering the possibility of net merging. As a secondary goal, the algorithm also tries to minimize the total wire length. Experimental results show that the algorithm is superior to many other methods.

7.2. Problem Formulation

Let C be the given circuit, $T = \{t_1, t_2, \ldots, t_l\}$ be the set of transistors in C, and $G = \{g_1, g_2, \ldots, g_n\}$ be the set of gate lines in C. Define a *node set* $P = \{p_1, p_2, \ldots, p_m\}$ to be a set of points p_i assigned between transistors in C. We assume that there are two "fictitious" nodes that correspond to power and ground connections. We have the following definitions:

A node set P is said to be *complete* if each transistor in C is connected to two nodes in P. Given a complete node set $P = \{p_1, p_2, \ldots, p_m\}$, we define the corresponding net–list $N = \{N_1, N_2, \ldots, N_m\}$ where N_i ($\subseteq G$) is the set of gate lines that are connected to node p_i. A complete node set is shown numbered and circled in Fig.7.1(a). The corresponding net–list is shown in Fig.7.1(b). In the final layout, each net is realized by a horizontal metal line which is connected to a drain, source, or gate of the transistors of the node. (See Fig.7.1(c).)

A *gate permutation* π is defined to be an ordering of the gate lines in C. Given a gate permutation π, we define the ***net interval*** for net N_k to be the interval $[l_k, r_k]$ where l_k and r_k are the positions of the leftmost and rightmost gate lines in net N_k, respectively. We define the *local density at column j*, denoted d_j, to be the number of nets that intersects gate line j and and the ***density d*** to be the maximum local density. Fig.7.1(c) and Fig.7.1(e) show two solutions with different gate permutations. In general, it is more convenient to represent these layout solutions by the corresponding abstract layout shown in Fig.7.1(d) and Fig.7.1(f), respectively. Nets are represented as horizontal line segments, and we use a "dot" to indicate where there is a transistor, a gate/output connection, or a diffusion run connection. The translation from layout solutions to abstract layout and vice versa are straight–forward and will not be given here. The net intervals are evident from the abstract layout shown in Fig.7.1(d) and Fig.7.1(f).

We note that the density is a function of the gate permutation π. For example, the density is 6 for the layout shown in Fig.7.1(d) while it is 4 for that shown in Fig.7.1(f). In general, the density is a *lower bound* on the number of tracks needed. Indeed, using the "Left Edge Algorithm" [HaSt71], for a given a net list N, and a gate permutation π with density d, we can lay out the nets in N in d horizontal tracks. Thus, to obtain an optimal layout (that uses the minimum number of tracks), we only need to find a gate permutation π that minimizes the density d. The "Left–Edge Algorithm" will then assign the nets in N to rows in the Gate Matrix. (We shall refer to an assignment of nets to rows in a Gate Matrix as a *row assignment*.) We can now restate the Gate Matrix layout problem as follows:

Given a circuit C, determine a net–list N and a gate permutation π such that the density d is minimized.

However, the above formulation is incomplete because it did not take into account some practical considerations that affect the final layout. In particular, it did not pay any attention to the routing of the diffusion runs, and some simplification steps can be incorporated.

Consider the layout of a group of transistors connected in series. An example is shown in Fig.7.2(a). Fig.7.2(b) shows the layout of three series transistors using two nets. Note that nets 1 and 4 are "external" nets. Fig.7.2(c) shows an alternative layout using only one net. Note, however, that in this layout only transistors a and c can be used to connect to the external nets. A set of transistors that are connected in series is called a ***series–net.*** An ***end–transistor*** of a series–net is either the leftmost or the rightmost transistor in that net. In general, the order in which the transistors in a series–net are connected is logically immaterial provided that an end–transistor is used to connect to the external net. Thus, it is possible and better to group these transistors as a single net. The order of the transistors will be determined by the gate permutation π. We shall use the term ***binding*** to refer to the assignment of the end–transistor to be connected to the external

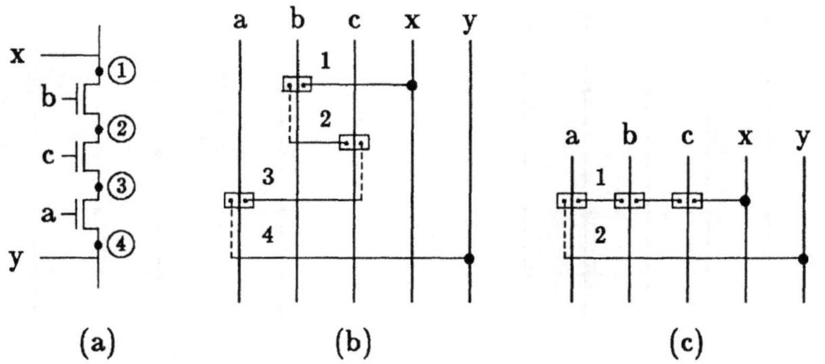

Fig. 7.2 Layout of a series–net. (a) Series–net. (b) A layout. (c) Alternative layout.

net. For example, for the series–net 3 in Fig.7.1(c), the current binding is valid since e is an end–transistor. It is also easy to see that by interchanging the binding to the other end–transistor that is connected to gate line e, the net interval for net 2 can be decreased from [1,6] to [3,6]. Notice that binding can only be considered after the gate permutation π is determined.

Consider nets 1 and 2 in Fig.7.3(a). It is quite clear that the short diffusion run to connect the net 2 to transistor b can be eliminated if we "merge" the two nets into a bigger net. The merged net is shown in Fig.7.3(b). However, nets 3 and 4 cannot be merged. Two nets can be *merged* (with respect to gate permutation π) if and only if their net intervals $[x_1, x_2]$ and $[x_2, x_3]$ are disjoint with the exception of the intersection at gate line x_2 and they share a transistor at column x_2. The merged net will have a net interval $[x_1, x_3]$. The advantages of merging are that it eliminates the diffusion run connection and also that it reduces the local density at column x_2 by 1. This can potentially reduce the density of the problem and hence the number of tracks in the layout. In Fig.7.3, the density goes down from 4 to 3 after merging.

Both binding and net merging affect the net intervals and hence should be applied before performing row assignment using the "Left–Edge Algorithm." In contrast, we now describe a

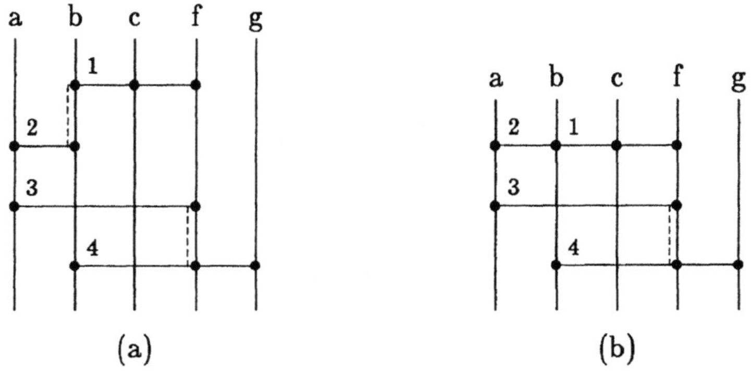

Fig. 7.3 Illustration of net merging. (a) Initial net intervals. (b) After merging.

practical consideration that is applied after row assignment is done. Consider the layout shown in Fig.7.4(a) obtained after applying the "Left–Edge Algorithm" to perform assignment of the nets to rows. We note that the diffusion run used to connect nets 2 and 5 can be realized. However, the diffusion run between nets 1 and 4 cannot be realized because there is a transistor along the path.

Given a row assignment, a diffusion run is said to be *realizable* if there is no transistor on its path. A row assignment is said to be *realizable* if every diffusion run is realizable. In [WiHW85], if a row assignment is not realizable, the rows of the Gate Matrix are permuted to obtain a realizable row assignment. In the example shown in Fig.7.4(a), a realizable row assignment can be derived by permuting the 2 *nd* and the 3 *th* rows so as to obtain the final layout shown in Fig.7.4(b).

It is well–known that the problem of minimizing the area in Gate Matrix layout is NP–complete [KaFu79]. In fact, the result in [KaFu79] showed that even finding an optimal solution for Step 2 (see below) is NP–complete. Hence, there is little hope of ever finding fast (polynomial time) algorithms for obtaining a layout with minimum area. Rather, most researches have focussed on designing fast heuristics for the problem. In [WiHW85], the

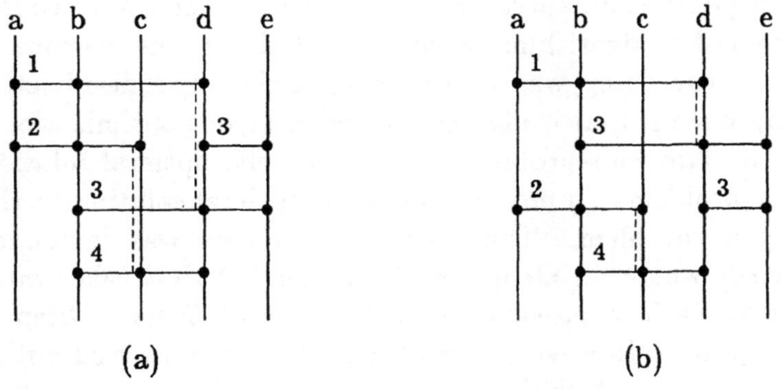

Fig. 7.4 Permutation of rows. (a) Initial row assignment. (b) After row rearrangement.

following general procedure for solving the Gate Matrix layout problem was proposed:

Step 1 : Determine the net–list N,
Step 2 : Determine gate permutation π (π is fixed for the remaining steps),
Step 3 : Identify end–transistors and process series–net,
Step 4 : Merge nets wherever possible,
Step 5 : Assign nets in N to row in the Gate Matrix,
Step 6 : Modify row assignment to obtain realizable row–assignment.

Earlier algorithms for Gate Matrix layout [Wing82, Wing83, Li83] address only Step 2. In these algorithms, the problem of determining a gate permutation in Step 2 was formulated as the problem of adding edges to a connection graph so that the augmented graph becomes an interval graph. The goal is to minimize the size of the largest clique in the interval graph, since that corresponds exactly to the density [Golu80]. To this end, various greedy heuristics were proposed [Wing82, Wing83, Li83, DeKL87]. The disadvantages of these algorithms are that they tend to produce only locally optimal solutions, and also that they operate at an abstract level without taking into account the practical considerations discussed above.

In [WiHW85], these practical considerations were considered and heuristic algorithms were presented for the various steps. However, each step was treated separately. The disadvantage of this approach is that it divides the global layout optimization problem into many sub–problems. Unfortunately, optimal solutions for the sub–problems do not guarantee an optimal solution to the global layout problem. The concept of D–nets was introduced in [HwFK86] where an algorithm for Steps 1–3 that takes into consideration various possible net–list representations in Step 2 was also proposed. However, Steps 2 and 3 are not carried out simultaneously in the algorithm; rather, when performing Step 2, information about series–nets and their external connections are taken into account.

7.3. Generalized Problem Formulation

We now present a generalization of the problem formulation of [WiHW85] which leads to an integrated approach to the Gate Matrix layout problem. The new features in the generalized formulation are:

(1) the use of the *D–net representation* and the concept of *dynamic binding* introduced by [HwFK86], and
(2) integration of Steps 1–5 so that they are considered simultaneously.

Thus, an algorithm for the generalized problem should consider the global optimization problem of permuting the gates so as to minimize the layout area, while at the same time take into consideration dynamic binding and merging.

We first describe the D–net representation. For a given circuit, there are many different net–list representations. In particular, in a series net with k transistors, depending on the gate permutation, there are k possible bindings of an end–transistor to an external node. In [WiHW85], bindings are determined in Step 1 while in Step 2 gate permutation is determined on the basis of the binding chosen in Step 1. Then in Step 3, the bindings are modified to take care of practical constraints while the gate permutation is held fixed. The disadvantage of this approach is that a poor choice of the bindings in Step 1 can lead to a poor choice of the gate permutation and hence to a poor overall layout.

In order to allow the binding to be determined *dynamically* (after the gate permutation is known) we use the following net–list representation. A net N_j is said to be a *dynamic* net (D–net) if it is an external net of some series–net N_k with at least two transistors. We shall denote this fact by having an entry $\{N_k\}$ in the net N_j to mean that an end–transistor of N_k needs to be bound to net N_j. However, this binding is not fixed but will be determined dynamically.

Consider the circuit schematic shown in Fig.7.5(a). In the fixed net–list representation of [WiHW85], the net–list is shown in Fig.7.5(b). Using the new formulation, we have the net–list shown in Fig.7.5(c), where, for example, N_3 is a series–net while N_2 is a

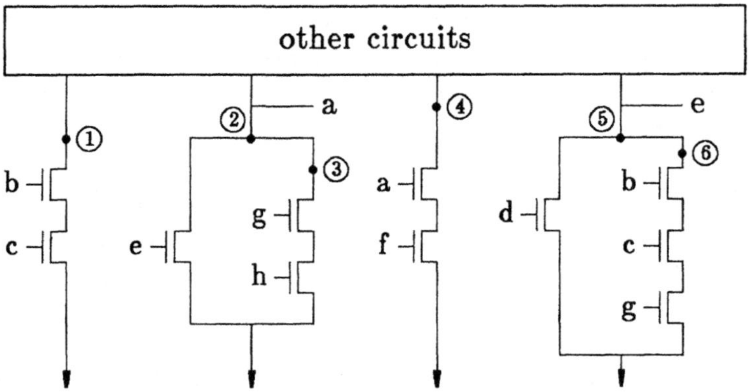

(a)

(b)	(c)
$N_1 = \{ \text{b, c} \}$	$N_1 = \{ \text{b, c} \}$
$N_2 = \{ \text{a, e, g} \}$	$N_2 = \{ \{N_3\}, \text{a, e} \}$
$N_3 = \{ \text{g, h} \}$	$N_3 = \{ \text{g, h} \}$
$N_4 = \{ \text{a, f} \}$	$N_4 = \{ \text{a, f} \}$
$N_5 = \{ \text{b, d, e} \}$	$N_5 = \{ \{N_6\}, \text{d, e} \}$
$N_6 = \{ \text{b, c, g} \}$	$N_6 = \{ \text{b, c, g} \}$

Fig. 7.5 Circuit schematic and net–list representations. (a) Circuit schematic. (b) Fixed net–list. (c) Dynamic net–list.

D–net. The entry $\{N_3\}$ in net N_2 means that one of the transistors in N_3, namely, either g or h, needs to be bound to net N_2. However, the choice is not finalized in Step 1; rather it varies as a function of the gate permutation, thus the name *dynamic binding*. We can now formulate the Generalized Gate Matrix Layout Problem as follows:

> Given a circuit schematic C, determine a dynamic net–list N, a gate permutation π, a set of bindings, and a set of merged nets so as to minimize the resulting density.

Note that in this formulation dynamic binding is explicitly

represented, and that Steps 1–5 are integrated into the problem statement instead of being stated as individual steps. The use of dynamic binding is very similar to the channel routing problem when physically or logically equivalent pins are present in the channel as was discussed in Chapter 5. The general strategy of delaying some design decisions so that these decisions can be incorporated into later design steps was also considered in [Leon86a].

We note that for the generalized formulation, the circuit schematic must be given. If the input is only the net–list (without the circuit schematic), then Steps 3–4 and 6 are not meaningful. The problem model then reduces to that considered in [Wing82, Wing83, Li83, DeKL87]. However, the simulated annealing algorithm to be presented in Section 7.5 will still work for this model.

7.4. Advantages of the Generalized Formulation

We now demonstrate the advantages of the generalized formulation over that of [WiHW85]. Consider the circuit shown in Fig.7.5(a). Using the approach of [WiHW85], a fixed net–list representation (shown in Fig.7.5(b)) is chosen in Step 1. In Step 2, the gates are permuted to minimize the density. It is obvious, by looking at gate b, that at least three tracks are required. Fig.7.6(b) shows an optimal gate permutation for Step 2. The final layout obtained after completing the other steps is shown in Fig.7.6(a). In particular, it is not difficult to see that once this gate permutation

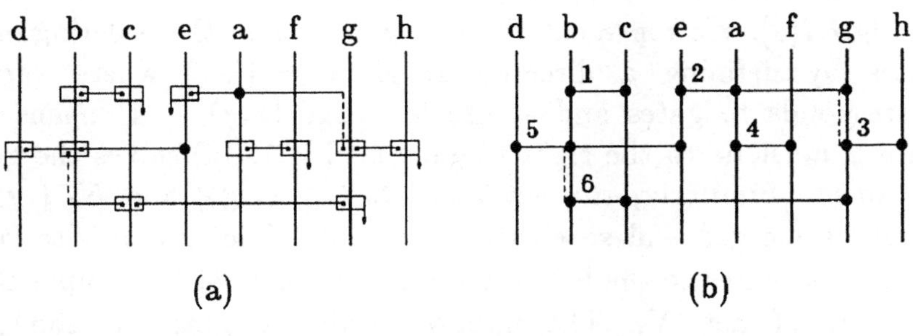

Fig. 7.6 Optimal solution using fixed net–list. (a) Actual layout. (b) Abstract layout.

is fixed to be that in Fig.7.6(b), a layout solution with two tracks cannot be derived in the subsequent steps. In fact, we shall show that for this problem, starting from *any* gate permutation for Step 2, no layout solution with two tracks can be derived in subsequent steps.

Lemma 7.1. Given the fixed net–list representation N for the circuit in Fig.7.5(a), a necessary condition to obtain a two track final layout is that (1) nets 2 and 3 must be merged, and (2) nets 5 and 6 must be merged.

Proof. Let $G(k)$ be the set of nets that intersects gate k, namely, $G(k) = \{N_j \mid k \in N_j\}$. Then $|G(k)|$ is a lower bound on the number of tracks needed for the layout. Now $G(\text{b})=\{1,5,6\}$. Thus, if nets 5 and 6 are not merged at gate line transistor b, then at least three tracks are required. Similarly, since $G(\text{g})=\{2,3,6\}$, nets 2 and 3 must also be merged at gate line g. □

Theorem 7.1. Given the fixed net–list representation N, and a gate permutation π obtained in Step 2. If π is fixed in subsequent steps, then a final layout with two tracks cannot be obtained.

Proof. Assume the contrary that π is a gate permutation that yields a two track final layout. By Lemma 7.1, the net pair N_5 ({b, d, e}) and N_6 ({b, c, g}) must be merged at gate b. Thus, either N_5 lies completely to the left of N_6 or vice versa. Without loss of generality, assume that N_5 is to the left of N_6 as illustrated in Fig.7.7(a). We represent these constraints on the ordering of the gates by defining a directed graph $G=(V,E)$ where vertices corresponds to gates and a directed edge $(x,y) \in E$ means that gate y must lie to the right of gate x. Fig.7.7(a) shows the graph obtained. Similarly, the net pair N_2 ({a, e, g}) and N_3 ({g, h}) must be merged and so either N_2 lies to left of N_3 or vice versa. Since gate e lies to the left of gate g, net N_2 must lie completely to the left of net N_3. The updated ordering graph is shown in Fig.7.7(b).

Consider the ordering of the subset {e, b, g, h} of gates. Fig.7.7(b) shows the partial layout where some of the nets are shown partially completed. From Fig.7.7(b) it is clear that at least three tracks are required. This is a contradiction. Therefore, no

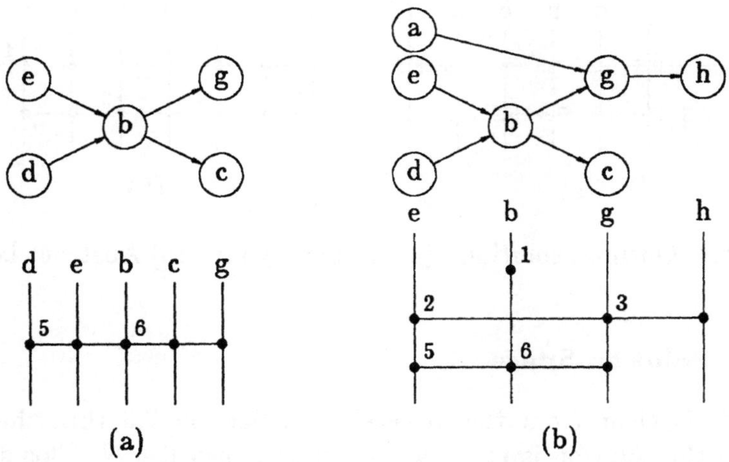

Fig. 7.7 Gate ordering graph G. (a) N_5 to the left of N_6. (b) N_5 to the left of N_6, N_2 to the left of N_3.

two track final layout exists. □

By Theorem 7.1, in the formulation of [WiHW85], the solution shown in Fig.7.6 is optimal. However, using the generalized formulation, we can obtain a solution with two tracks as shown in Fig.7.8(b). Fig.7.8(a) shows the final layout. Notice that in this optimal solution, nets N_5 and N_6 are merged at gate g while nets N_2 and N_3 are merged at gate h (as a consequence of the concept of dynamic binding). In fact, this problem was tested on an implementation of the simulated annealing algorithm, and this optimal solution using two tracks was obtained in less than 1 sec. of CPU time.

7.5. The Simulated Annealing Method

In this section, we describe in detail a simulated annealing layout algorithm [Leon86b] that solves the generalized Gate Matrix layout problem by simultaneously considering Steps 1–5.

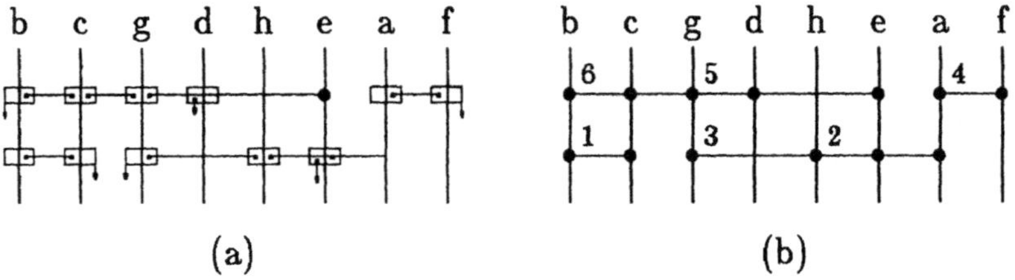

Fig. 7.8 Optimal solution. (a) Actual layout. (b) Abstract layout.

7.5.1. Solution Space

It is clear from the discussion in Section 7.3 that the first step in the solution is to obtain a dynamic net–list N. This step is straight–forward if we are given the circuit schematic C. Thus, in what follows, we shall assume that N has been obtained and in the discussion we shall omit the net–list N whenever there is no confusion.

In general, any *layout solution* is completely described by specifying (1) the gate permutation π, (2) the dynamic binding, (3) the set of merged nets, and (4) the row assignment for each net. However, we shall argue that (1)–(2) are sufficient to represent any layout solutions. Given the gate permutation π and the dynamic bindings, the set of net intervals is completely defined. Thus, the set of merged nets can be uniquely identified. To speed up this computation, a list of all pairs of nets that share a transistor (and are therefore potentially mergable) is maintained and the algorithm is straight–forward. Then the assignment of nets to rows can be done optimally in d tracks using the "Left–Edge Algorithm" [HaSt71]. Thus, the problem of minimizing the number of tracks in the layout solution is equivalent to that of minimizing the density given (1)–(2).

For the simulated annealing layout algorithm, a *layout solution* is specified by (1) the *gate permutation* π and (2) the *dynamic bindings*. The algorithm dynamically keeps track of (1), (2), and the local densities of the intermediate solutions. Thus, we have the ability to integrate Steps 1–5 while at the same time have a very compact solution representation that is suitable for simulated annealing.

7.5.2. Neighboring Solutions

Given a layout solution, namely, the gate permutation and the dynamic bindings, we define two types of *moves*:

M1. Interchange the positions of two of the gates in the gate permutation.
M2. Interchange the binding of a D–net.

For a move of type M2, the binding of the chosen D–net is reversed. The gate permutation and the bindings of all other D–nets are unchanged. For a move of type M1, the gate permutation is perturbed and hence the following might be affected: (1) nets that use any of the permuted gates, (2) bindings for some D–nets, and (3) the set of merged nets. Fixed nets are easily updated. For dynamic nets that are affected, the bindings need to be updated. To this end, if one binding is obviously superior to the other, the better binding is always chosen. For example, in Fig.7.9(a) if the transistor with gate line a was bound to net N_1, the net interval for N_1 would be [1,6] whereas if the transistor with gate line a was used the net interval would be [3,6] ($\subseteq$ [1,6]). Thus, binding d is clearly superior and will therefore be chosen. On the other hand, in the case that it is not clear which of the bindings is a better one, the choice is guided by the simulated annealing procedure. In the example in Fig.7.9(b), the net interval for net N_1 corresponding to the two possible bindings (either c or e) are [1,4] and [2,5],

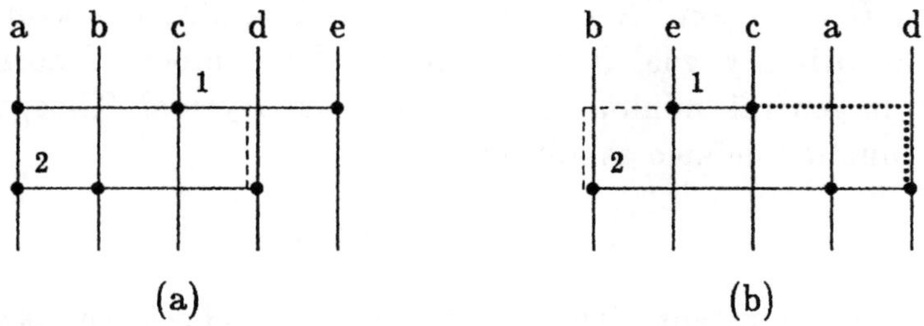

(a) (b)

Fig. 7.9 Binding a dynamic net $N_1 = \{ \{N_2\}, \text{c}, \text{e}\}$ to series–net $N_2 = \{\text{a, b, d}\}$. (a) Transistor d bound to net N_1. (b) Transistor c bound to net N_1.

respectively. Since neither is better than the other, we shall let the choice be determined by the moves in the simulated annealing process.

After obtaining the improved net definition, the new set of merged nets are computed. In particular, two nets with corresponding intervals $[x_1, x_2]$ and $[x_2, x_3]$ and sharing the same transistor at x_2 can be merged. After merging, their intervals are replaced by one interval $[x_1, x_3]$. Note that the local density at column x_2 is decreased by one after the merge.

Clearly, given a layout solution, it is possible to go to any other layout solution via a sequence of moves of type M1 and M2. Two layout solutions are said to be *neighbors* if one can be obtained from the other by a move. In the simulated annealing algorithm these moves are generated at random. Clearly, we should bias the generation of moves so that more moves of type M1 than moves of type M2 will be attempted since they allow more global rearrangements of the nets. Furthermore, efficient procedures are used to process each move so that only nets affected by the move need to be modified.

7.5.3. Cost Function

Let n be the number of gates, m be the number of nets, $D = \sum_{j=1}^{n} d_j$, and W be the sum of the lengths of the nets. It is easy to see that $D = W+m$. Recall that for the Gate Matrix layout problem, the primary goal is to minimize the number of horizontal tracks. In general, minimizing W is a secondary goal. Thus, we use a cost function defined as follows:

$$C = d^2 + \frac{\lambda}{n} D^2,$$

where λ is a constant. The first factor in C reflects the fact that the primary goal is to minimize d. In the second factor, we use D to approximate W. Thus, this factor is approximately equal to the average of the sum of squares of the wire length and it reflects the secondary goal. The value of λ is used to control the relative importance of d and W. Thus, we should have $\lambda \leq 1$. Experimental results indicate that 0.5 is a good choice as the default value of λ.

7.5.4. Annealing Schedule

The temperature schedule is of the form $T_k = r \cdot T_{k-1}$, k=1,2,... . Experimental results indicate that setting r = 0.8 produces very good results. In determining the initial temperature T_0, we see what happens if a move causes the density to increase from d to $d+K$. The corresponding in C is $\Delta \simeq 2Kd+K^2$ and the probability of accepting this move is $e^{-\Delta/T}$. Thus, for there to be a reasonable probability of acceptance at T_0, we should have $T_0 = O(d)$. In the current implementation, T_0 is of the form $\lambda_T \cdot m$ where m is the number of nets and is an upper bound of d.

7.6. Experimental Results

The simulated annealing algorithm for Gate Matrix layout was implemented in Pascal on a PYRAMID computer. The following problems from [Wing82, Wing83, Li83, WiHW85] were used to compare the performance of the algorithm with that of other algorithms.

GM1	Figure 3 of [Li83]
GM2	Figure 10 of [Wing82]
GM3	Figure 3 of [Wing83]
GM4	Figure 14 of [WiHW85]
GM5	Figure 10 of [WiHW85]

The results obtained by the simulated annealing layout algorithm are shown in Table 7.1. In some instances (GM1–GM3), it is not possible to obtain the circuit schematic, and hence the concept of D–nets, dynamic binding and merging cannot be applied. For these examples, the generalized formulation becomes the same as that of [Wing82] and [Wing83]. Table 7.1 shows that the simulated annealing algorithm also performs optimally for all these problems. Thus, these results indicate that the algorithm also produces good results for the more restricted problem formulation.

For GM4 and GM5, the simulated annealing layout algorithm performs better than that of [WiHW85]. For GM4, the number of tracks used was reduced from 5 to 4. For a large

Problem Specification				Number of Tracks	
Desc.	No. of Gates	No. of Nets	No. of Trans	Other Methods	Simulated Annealing
GM1	10	11	?	4	4
GM2	21	18	?	4	4
GM3	25	16	?	8	8
GM4	8	8	12	5	4
GM5	71	99	306	40	29

Table 7.1

problem, GM5, the improvement is drastic, from 40 tracks[1] to 29 tracks – an improvement of 27.5%. The abstract layout is shown in Fig.7.10.

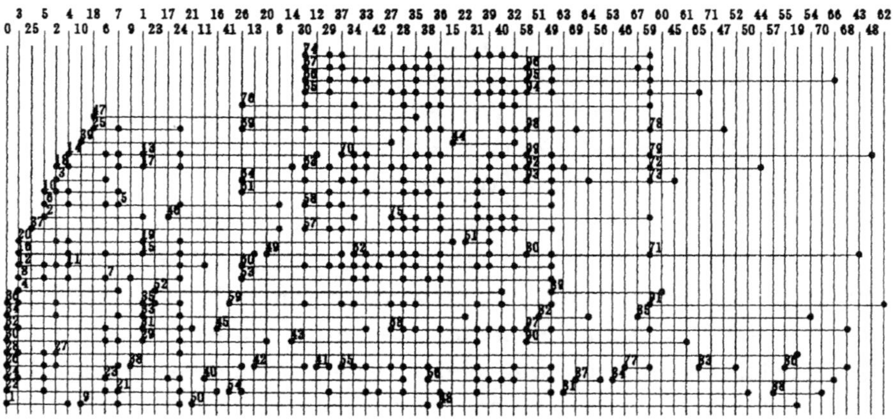

Fig. 7.10 Final result for GM.5 with 71 gates, 131 nets, and 306 transistors. The solution uses 29 tracks.

[1] In [WiHW85], it was reported that the solution uses 42 tracks. However, a 40 track solution has since then been reported.

As a further test of the algorithm, problems GM1 and GM2 were used again to measure the contribution of the integration of Steps 1–5 to the performance of the algorithm. To this end, the algorithm is applied to GM4 and GM5 with the contribution of the dynamic net–list representation factored out by using the same fixed net–list representation that was used in [WiHW85]. Namely, only the method of integrating Steps 1–5 is used. For GM4, the algorithm produced the 4 track solution as expected. The final layout is shown in Fig.7.11. For GM5, the result is more spectacular in that a 29 track solution was obtained. Thus, we can conclude that while the generalization to dynamic net–list is important as proven in Section 7.4, the main contribution to the power of the method comes from the flexibilities made possible by integrating Steps 1–5.

The running time of the algorithm depends on the number of gates. For problems GM1–GM4, the running times were 2.0, 61.2, 3.6, and 2.1 seconds, respectively. For the large problem, the total time taken for the annealing process was 13.6 minutes. However, even after 3 seconds, a solution with 33 tracks has already been found. In fact, this behavior is generally true of many simulated annealing algorithms [LeLi85, LeWL85, WoLL86] that try to minimize density. Near optimal solutions are obtained fairly quickly, and then it gets progressively harder to decrease the

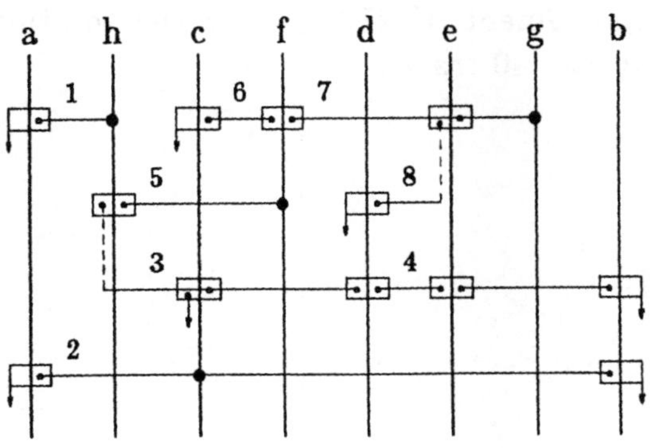

Fig. 7.11 A 4 track layout solution for GM4 using fixed net–list representation.

number of tracks, even though the average cost function is slowly decreasing. In general, one can trade off computing time against the quality of the solution. For example, one can set a time limit, and then settle for the best solution found so far. As indicated by GM5, the solution thus obtained is still very good.

7.7. Concluding Remarks

The Gate Matrix layout problem is known to be an NP-complete problem. Most algorithms for automatic Gate Matrix layout construct the solution in a number of separate steps. In this chapter, we presented a generalized formulation of the problem that unifies the sub-steps so that they are considered simultaneously. We showed that the generalized formulation is indeed superior to that of [WiHW85].

We also presented an algorithm for solving the problem in the generalized formulation. The algorithm is able to minimize the layout area and total wire length by taking advantage of dynamic binding for D-nets and net-merging. Experimental results indicate that the method produces layouts that are good in terms of total area and the sum of the wire length. Solutions of some benchmark problems demonstrate substantial improvements in the number of tracks required for layout. For a big example with 71 gates, 99 nets, and 305 transistors, the algorithm obtained a solution with 29 tracks, an improvement of 27.5% over the solution obtained by [WiHW85] that uses 40 tracks.

CHAPTER 8

ARRAY OPTIMIZATION

8.1. Introduction

In this chapter, we consider the problem of minimizing layout area of logic arrays. Most of the previous research efforts [DeSa83, HaNS82, WoLL87] on the subject have been concentrated on reducing layout area of Programmable Logic Arrays (PLA), a class of widely used logic arrays which are two–level implementations of logic functions. In general, multi–level implementations such as Gate Matrix, Weinberger Array, SLA, etc., can substantially reduce the total layout area and improve the speed performance of a circuit. Existing minimization methods that were primarily designed for PLA's are no longer effective for these more general logic arrays. A general formulation of an array optimization problem for efficient layout of multi–level logic circuits was first proposed in [DeNe86] and further studied in [WoLi87b]. The general optimization problem includes compaction of Gate Matrix layouts, SLA's, Weinberger Arrays, and for multiple folding of PLA's as special cases. We shall present in this chapter an algorithm proposed in [WoLi87b] for the array optimization problem. As in [DeNe86], the approach of [WoLi87] is also based on the method of simulated annealing. Despite the fact that both of these approaches use the same general optimization technique, the two algorithms are quite different. The major difference between the algorithms [DeNe86] and [WoLi87] is in the solution spaces they have chosen. The algorithm in [DeNe86] uses a solution space that includes both valid and invalid solutions. For their solution representation, it is crucial that many invalid solutions are included in the solution space since generating moves that visit only valid solutions is computationally expensive. Consequently, invalid solutions are needed to act as intermediate solutions between neighboring valid solutions. On the other hand, [WoLi87b] took a different approach in designing the solution space. The solution space in [WoLi87b] has the following features: (1) every solution in the solution space corresponds to a valid solution; (2) it only consists of a subset of all the possible valid solutions; (3) all optimal solutions† are represented in the solution space; and (4) it is very easy (computationally) to move from one solution to a neighboring solution. Note that a smaller solution space in general makes the search process more effective.

This chapter is organized as follows. A precise formulation of the array optimization problem is given in Section 8.2. Section 8.3 contains definitions which will be needed in later sections. Section 8.4 describes in detail the array optimization algorithm proposed in [WoLi87b]. Experimental results are presented in Section 8.5. Finally, some concluding remarks are given in Section 8.6.

8.2. The Array Optimization Problem

We now introduce the array optimization problem. In a given (unfolded) array as shown in Fig.8.1(a), cells of varying sizes are situated at grid points. These cells may represent single transistors or collections of transistors and interconnects. Rows and columns in the array correspond to different signal nets. The signal nets in the columns are referred to as *vertical nets*, and the signal nets in the rows are referred to as *horizontal nets*. Each cell in the array is connected to a horizontal net and a vertical net. It is possible to reduce the area of the array by allowing different vertical nets to share a column and different horizontal nets to share a row. Fig.8.1(b) shows a folded version of the array in Fig.8.1(a). Note that the order of placement of the cells in a horizontal or vertical net is not important.

The *width* of a column in an array (folded or unfolded) is defined to be the maximum cell width among all the cells in that column. The *height* of a row is defined to be the maximum cell height among all the cells in that row. The *width* of an array is the sum of the widths of the columns. The *height* of an array is the

† Strictly speaking, there are some optimal solutions which are not represented in the solution space. But these unrepresented solutions are just minor rearrangements of the optimal solutions (with the same cost) represented in the solution space.

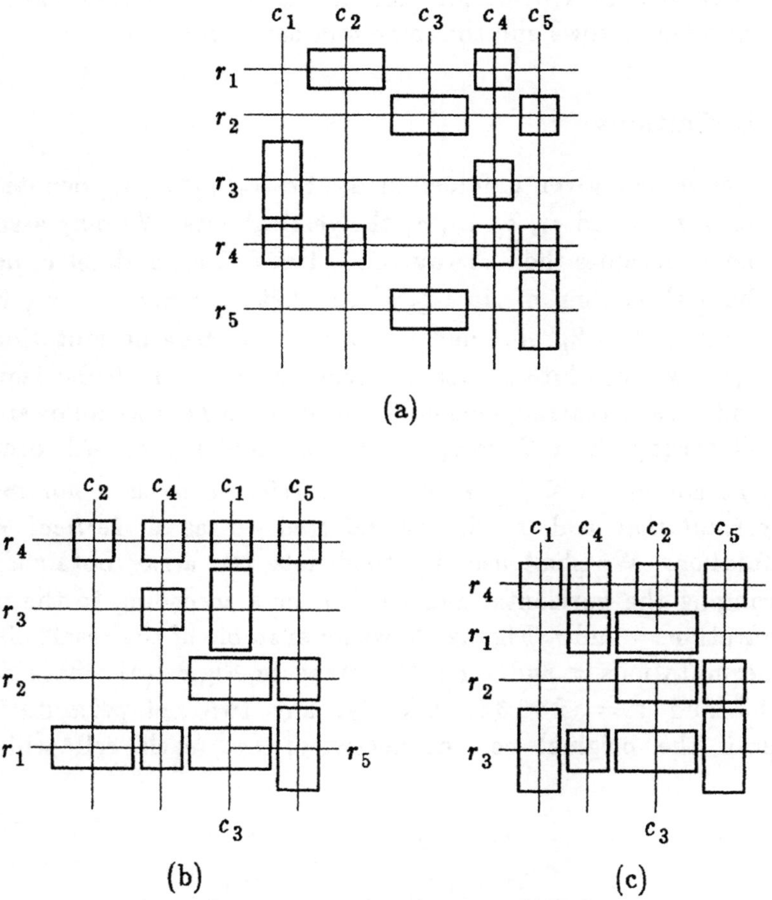

Fig. 8.1 Array folding. (a) An unfolded array. (b) A folded version of the array in (a). (c) Another folded version of the array in (a).

sum of the heights of the rows. The *area* of an array is the product of the width and the height of the array. (We shall use $AREA(A)$ to denote the area of an array A.) The objective of the array optimization problem is to find a folding of the rows and the columns of a given array such that the area of the folded array is minimized. Note that because of the varying cell sizes, simply minimizing the product of the number of rows and the number of columns in general does not yield a minimum area folding solution. For example, Fig.8.1(c) shows another folding solution of the array in Fig.8.1(a). The area of the array in Fig.8.1(c) is actually smaller than that in Fig.8.1(b) despite the fact that both arrays have the same number of rows and the same number of columns.

8.3. Definitions

Let A be a given unfolded array. Let $r_1, r_2, \ldots, r_m$ denote the horizontal nets and $c_1, c_2, \ldots, c_n$ the vertical nets. We may assume that net r_i occupies the i *th* row of A, $1 \leq i \leq m$, and net c_j occupies the j *th* column of A, $1 \leq j \leq n$. Let $\pi = \pi_1\pi_2 \cdots \pi_m$ be a permutation of $\{1,2,\ldots,m\}$ and $\tau = \tau_1\tau_2 \cdots \tau_n$ be a permutation of $\{1,2,\ldots,n\}$. We can interpret π as a rearrangement of the horizontal nets and τ as a rearrangement of the vertical nets as follows: net r_{π_i} will occupy the i *th* row, $1 \leq i \leq m$, and net c_{τ_j} will occupy the j *th* column, $1 \leq j \leq n$. π will be referred to as a *horizontal net-permutation* and τ will be referred to as a *vertical net-permutation.* We shall use $A_{\pi\tau}$ to denote the array obtained by rearranging the horizontal and vertical nets according to the net-permutations π and τ. Fig.8.2 shows an example of the result of the net-permutations π and τ for the array in Fig.8.1(a) where $\pi = 43215$ and $\tau = 24135$. Clearly, any two net-permutations preserve the original cell connectivity; *i.e.* each cell will be

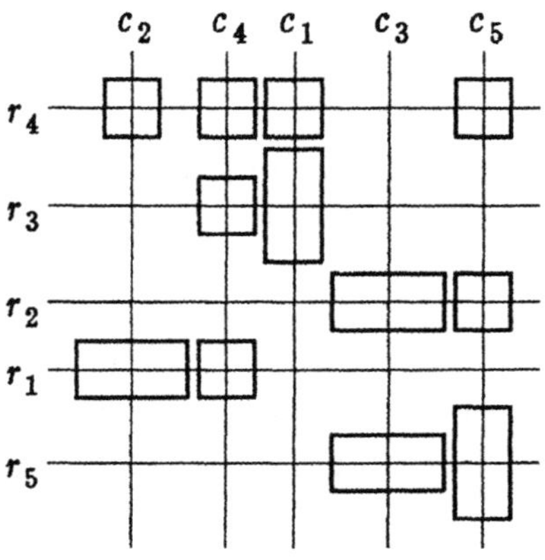

Fig. 8.2 Result of horizontal and vertical net-permutations for the array in Fig. 8.1(a).

connected to the same horizontal and vertical nets. A *segment* of a permutation τ is a set of the form $\{\tau_i,\tau_{i+1},...,\tau_{i+j}\}$. A partition $\Omega = \{S_1,S_2,...,S_p\}$ of $\{1,2,...,n\}$ where all the S_i's are segments of τ is called a *partition* of the permutation τ. For example, let $\tau = \tau_1\tau_2\tau_3\tau_4\tau_5\tau_6 = 2\ 1\ 6\ 4\ 3\ 5$. Let $\Omega = \{S_1,S_2,S_3\}$ with $S_1 = \{2,1,6\}$, $S_2 = \{4\}$, and $S_3 = \{3,5\}$. According to the definitions, S_1, S_2, and S_3 are segments of τ, and Ω is a partition of τ.

8.4. The Array Optimization Algorithm

In this section, we present the simulated annealing array optimization algorithm of [WoLi87]. We show how to represent folding solutions by the set of all triples (π,τ,Ω), where π and τ are two net–permutations, and Ω is a partition of τ. For an important special case in which the cells are of uniform size, we show that the solution space can be restricted essentially to the set of all pairs of net–permutations (π,τ) for most arrays. The algorithm depends on a special procedure to obtain a folding solution from a given triple (π,τ,Ω). The procedure consists of two phases. In the first phase, the columns are folded by an algorithm called COLUMN–FOLD. In the second phase, the rows of the column–folded array obtained in the first phase are folded by an algorithm called ROW–FOLD.

8.4.1. The Algorithm COLUMN–FOLD

We now describe the algorithm COLUMN–FOLD that folds the columns of $A_{\pi\tau}$. For each column j of $A_{\pi\tau}$, $1 \le j \le n$, we define a vertical interval $I_j = [a_j,b_j]$ where a_j and b_j are the positions of the centers of the topmost and the bottommost cells in the column. Fig.8.3 illustrates the vertical intervals of an array. Two vertical intervals are said to be *disjoint* if they do not overlap. Clearly, the lengths and positions of these intervals depend very much on the horizontal net–permutation τ which specifies the relative ordering of the cells in each column. The *width* of I_j is defined to be the width of column j of $A_{\pi\tau}$. Since column j of $A_{\pi\tau}$ corresponds to the vertical net c_{τ_j}, interval I_j also corresponds to the vertical net c_{τ_j}, $1 \le j \le n$. Let C be a set of pairwise disjoint vertical intervals. The *width* of C is defined to be the maximum of the widths of the intervals in C. It is clear that all the vertical nets corresponding to the intervals in C can be placed in the same

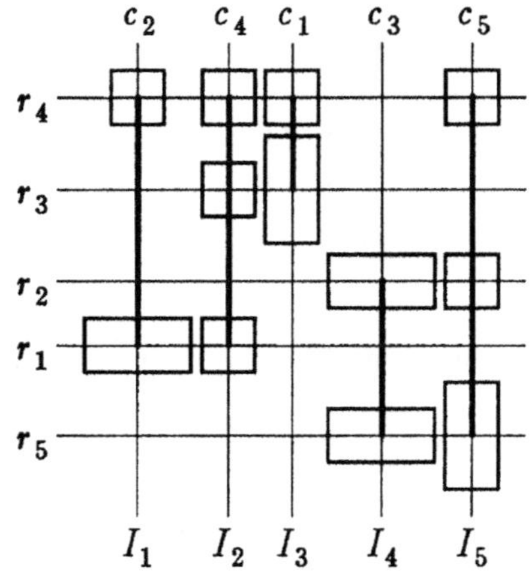

Fig. 8.3 Vertical intervals.

column, and the width of the resultant folded column is simply the width of C.

An *interval—segment* is a set of consecutive vertical intervals $\{I_i, I_{i+1}, ..., I_{i+j}\}$ where $\{r_i, r_{i+1}, ..., r_{i+j}\}$ is a segment of Ω. Clearly, the partition Ω induces a partition of the vertical intervals into interval–segments. We shall interpret the partition Ω as an additional restriction on whether some of the vertical nets can be placed in the same column. In fact, the restriction is that two vertical nets cannot be placed in the same column if their corresponding intervals I_i and I_j belong to different interval–segments. Let C be a set of disjoint vertical intervals from the same interval–segment. An interval $I_j \notin C$ is said to be *disjoint* from C if: (1) I_j is pairwise disjoint from all the intervals in C; and (2) I_j and all the intervals in C are in the same interval–segment.

The algorithm COLUMN–FOLD constructs a sequence of sets C_1, C_2, ..., C_{n_f} where each C_i is a set of pairwise disjoint vertical intervals that are in the same interval–segment, and all the C_i's together form a partition of $\{I_1, I_2, ..., I_n\}$. We can then fold the columns of $A_{\pi r}$ by placing in the i *th* column all the vertical nets corresponding to the intervals in C_i, $1 \leq i \leq n_f$. We shall use $A_{\pi r}^{C_\Omega}$ to denote the $m \times n_f$ column–folded array. The algorithm is shown in Fig.8.4. Effectively, the algorithm processes the intervals sequentially according to the order $I_1, I_2, \ldots, I_n$. At every iteration, we try to place interval I_i in one of the sets $C_1, C_2, \ldots, C_{n_f}$. If I_i is not disjoint from C_{n_f}, then we place I_i in a new set C_{n_f+1}. Otherwise we look for a set in which I_i is to be placed. C_{n_f} is a candidate. For $k < n_f$, C_k is a candidate if I_i is disjoint from C_k and the width of I_i is less than or equal to the width of C_k. We then place I_i into C_s where s is the smallest index such that all C_s, $C_{s+1}, \ldots, C_{n_f}$ are candidates.

The algorithm COLUMN–FOLD usually does not yield a column–folded array with minimum width. In fact, an extremely compact column–folded array in general provides less flexibility for folding the rows in the second phase. COLUMN–FOLD produces a column–folded array which usually allows ROW–FOLD to further reduce significantly the area of the array.

Algorithm COLUMN–FOLD

1. $C_1 \leftarrow \{I_1\}$; $n_f \leftarrow 1$; $i \leftarrow 2$.
2. If I_i is not disjoint from C_{n_f}, then $\{n_f \leftarrow n_f + 1;$ $s \leftarrow n_f$; $C_{n_f} \leftarrow \emptyset$; go to 5$\}$.
3. Find the smallest index j such that I_i is disjoint from C_j, $C_{j+1}, \ldots, C_{n_f-1}$, and the width of $I_i \leq$ the width of C_k, for all $k = j, j+1, \ldots, n_f-1$.
4. If such index j is defined, then $s \leftarrow j$, else $s \leftarrow n_f$.
5. $C_s \leftarrow C_s \cup \{I_i\}$; $i \leftarrow i + 1$.
6. If $i > n$, then STOP, else go to 2.

Fig. 8.4 Algorithm COLUMN–FOLD.

8.4.2. The Algorithm ROW–FOLD

We now describe the algorithm ROW–FOLD that folds the rows of $A_{\pi r}^{C_\Omega}$. Similar to the definition of vertical intervals, we define a set of horizontal intervals for the rows in $A_{\pi r}^{C_\Omega}$. For row i in $A_{\pi r}^{C_\Omega}$, $1 \leq i \leq m$, we define a horizontal interval $J_i = [g_i, h_i]$ where g_i and h_i are the positions of the centers of the leftmost and the rightmost cells in the row. Fig.8.5 illustrates the horizontal intervals of a column–folded array. The *height* of J_i is defined to be the height of row i. We can apply essentially the same algorithm for packing vertical intervals presented in Section 8.4.1 to pack the horizontal intervals. The only difference is that there is no restriction imposed by a partition of permutation such as Ω in the case of column–folding. Let R be a set of pairwise disjoint horizontal intervals. The *height* of R is defined as the maximum of the heights of the intervals in R. A horizontal interval J_i is *disjoint* from R if J_i is pairwise disjoint from all the intervals in R. The algorithm ROW–FOLD constructs a sequence of sets $R_1, R_2, ..., R_{m_f}$ where each R_i is a set of pairwise disjoint horizontal intervals, and all the R_i's together form a partition of $\{J_1, J_2, ..., J_m\}$. We can then fold the rows of $A_{\pi r}$ by placing all the horizontal nets corresponding to the intervals in R_i, $1 \leq i \leq m_f$, on the i *th* row. We shall use $A_{\pi r}^{C_\Omega R}$ to denote the $m_f \times n_f$ column– and row–folded array. The algorithm is shown in Fig.8.6.

It should be pointed out that $A_{\pi r}^{C_\Omega R}$ is indeed a folding solution of A. This is not totally obvious. It is true that any partition of the vertical intervals $\{I_1, I_2, ..., I_n\}$ into sets of pairwise disjoint

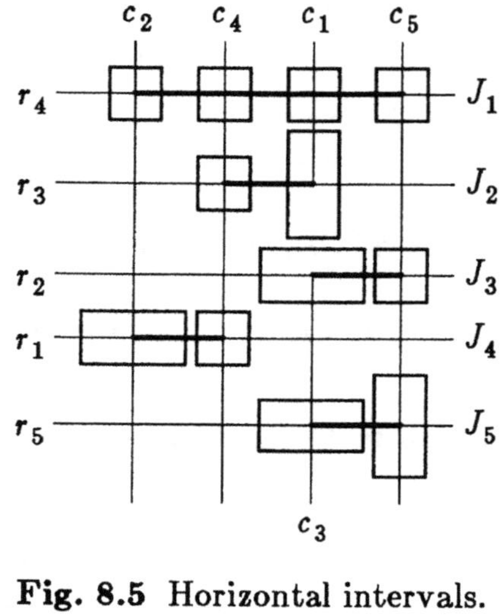

Fig. 8.5 Horizontal intervals.

Algorithm ROW–FOLD

1. $R_1 \leftarrow \{J_1\}$; $m_f \leftarrow 1$; $i \leftarrow 2$.
2. If J_i is not disjoint from R_{m_f}, then
   $\{m_f \leftarrow m_f + 1;\ s \leftarrow m_f;\ R_{m_f} \leftarrow \emptyset$; go to 5$\}$.
3. Find the smallest index j such that J_i is disjoint from $R_j, R_{j+1}, ..., R_{m_f-1}$, and the height of J_i $\leq$ the height of R_k, for all $k = j, j+1, ..., m_f-1$.
4. If such index j is defined, then $s \leftarrow j$, else $s \leftarrow m_f$.
5. $R_s \leftarrow R_s \cup \{J_i\}$; $i \leftarrow i + 1$.
6. If $i > n$, then STOP, else go to 2.

Fig. 8.6 Algorithm ROW–FOLD.

intervals corresponds to a column folding of A. But once we have folded the columns and obtained the set of horizontal intervals $\{J_1, J_2, ..., J_m\}$, it is no longer true that any partition of $\{J_1, J_2, ..., J_m\}$ into sets of pairwise disjoint intervals always corresponds to a folding solution of A. We have the following lemma.

Lemma 8.1. Given a pair of net–permutations π and τ, and a partition Ω of τ, application of the algorithm COLUMN–FOLD followed by the algorithm ROW–FOLD always yields a folding solution of A.

Proof. We only need to show that there are no *overlapping vertical nets* (*i.e.* overlapping vertical intervals in the same column) and no *overlapping horizontal nets* (*i.e.* overlapping horizontal intervals in the same row) in $A_{\pi\tau}^{C_\Omega R}$. It is clear that there are no overlapping horizontal nets because they are folded by ROW–FOLD which does not allow overlapping horizontal intervals to be placed on the same row. For the case of vertical nets, we first observe that there are no overlapping vertical nets in $A_{\pi\tau}^{C_\Omega}$ because COLUMN–FOLD does not allow overlapping vertical intervals to be placed on the same column. We further observe that column i of $A_{\pi\tau}^{C_\Omega}$ and column i of $A_{\pi\tau}^{C_\Omega R}$ contain the same set of cells (and hence the same set of nets), $1 \leq i \leq n$. Therefore, to show that there are no overlapping vertical nets in $A_{\pi\tau}^{C_\Omega R}$, it suffices to show that the relative ordering of the cells in column i of $A_{\pi r}^{C_\Omega}$ and the relative ordering of the cells in column i of $A_{\pi r}^{C_\Omega R}$ are the same, $1 \leq i \leq n$. (Hence, ROW–FOLD only "shrinks" the vertical intervals of $A_{\pi r}^{C_\Omega}$.) Let C_1 and C_2 be two consecutive cells in a column of $A_{\pi r}^{C_\Omega}$. We may assume that C_1 is above C_2. Let h_1 and h_2 be the horizontal nets that are connected to C_1 and C_2, respectively. Note that in $A_{\pi r}^{C_\Omega}$ the row containing h_1 is above the row containing h_2. Clearly, the two horizontal intervals corresponding to the two nets overlap. We observe that ROW–FOLD will neither place the two nets on the same row nor place h_2 in a row which is above the row containing h_1. Therefore, the relative ordering of C_1 and C_2 is preserved. □

8.4.3. The Solution Space

A folding solution is said to be *horizontally compact* if it satisfies the following conditions: (1) every interval I in column i either overlaps with an interval in column $i-1$ or the width of I is greater than the width of column $i-1$, for all $i \geq 2$; and (2) there exists an interval in column i which overlaps with an interval in column $i-1$, for all $i \geq 2$.

A folding solution is said to be *vertically compact* if it satisfies the following conditions: (1) every interval J in row j either overlaps with an interval in row $j-1$ or the height of J is greater than the height of row $j-1$, for all $j \geq 2$; and (2) there exists an interval in row j which overlaps with an interval in row $j-1$, for all $j \geq 2$.

A folding solution is said to be *compact* if it is both horizontally compact and vertically compact.

Lemma 8.2. Given any folding solution S, there exists a compact folding solution S_{hv} such that $AREA(S_{hv}) \leq AREA(S)$.

Proof. We define two operations *MERGE* and *SHIFT* that can be used to compact S. These operations are defined as follows:

(1) *MERGE*: Remove all the vertical (horizontal) intervals in column $i+1$ (row $i+1$) and place them in column i (row i) if the intervals in these two columns (rows) are pairwise

disjoint. Delete the empty column $i+1$ (row $i+1$) afterwards.

(2) *SHIFT*: Remove a vertical (horizontal) interval I from column i (row i) and place it in column $i-1$ (row $i-1$) if: (i) I and all the intervals in column $i-1$ (row $i-1$) are disjoint; and (ii) the width (height) of I is less than or equal to the width (height) of column i (row i).

By repeated applications of *MERGE* and *SHIFT*, we can obtain a compact folding solution S_{hv}. In fact, since both *MERGE* and *SHIFT* move at least one vertical (horizontal) interval from a column (row) to another column (row) with a smaller index, S_{hv} can be obtained by at most $m(m-1)/2 + n(n-1)/2$ operations. Since both types of operations never increase the area of the array, we have $AREA(S_{hv}) \leq AREA(S)$. □

We have shown that for any given pair of net–permutations π and τ, and a partition Ω of τ, we can construct a folding solution $A_{\pi\tau}^{C_\Omega R}$. Even though the set of all possible $A_{\pi\tau}^{C_\Omega R}$ in general does not contain all possible folding solutions, the following theorem shows that we would not miss any "good" solutions if we only restrict our search effort on the set of all possible $A_{\pi\tau}^{C_\Omega R}$.

Theorem 8.1. Let A be an unfolded array with variable cell sizes. Given any folding solution S, there exists a pair of net–permutations π and τ, and a partition Ω of τ, such that $AREA(A_{\pi\tau}^{C_\Omega R}) \leq AREA(S)$. Moreover, we have $|\Omega|^\dagger \leq \min\{m, n\}$.

Proof. Given a folding solution S, by Lemma 8.2, we can obtain a compact folding solution S_{hv}. We now show how to construct an unfolded array $A_{\pi\tau}$ by unpacking the horizontal and vertical intervals of S_{hv}. For each i, let $J_1^{(i)}, J_2^{(i)}, \ldots, J_{k_i}^{(i)}$ be the set of all horizontal intervals (k_i is the total number of intervals) in row i sorted according to the positions of their left end points, and let p_i be the total number of horizontal intervals in row 1, 2, ..., $i-1$; *i.e.* $p_i = \sum_{j=1}^{i-1} k_j$. We unpack the horizontal intervals by expanding each row of S_{hv} into a set of consecutive rows such that each row

† The number of segments in Ω.

contains exactly one interval. Let S_h denote the array we obtain after unpacking all the horizontal intervals. For each row i of S_{hv}, the k_i consecutive rows p_i+1, p_i+2, ..., p_i+k_i in S_h contain the k_i intervals $J_1^{(i)}, J_2^{(i)}, \ldots, J_{k_i}^{(i)}$. The assignments of the intervals to these rows should be such that we will obtain S_{hv} when we apply ROW–FOLD to S_h. Since S_{hv} is a compact folding solution, for each row i of S_{hv} there exists an interval J in this row such that it overlaps with an interval in row $i-1$. We assign this interval J to row p_i+1 of S_h and then arbitrarily assign the other intervals in row i of S_{hv} to row p_i+2, p_i+3, ..., p_i+k_i of S_h. Now using the fact that S_h is obtained from a horizontally compact solution (S_{hv}), it is not difficult to see that we will obtain S_{hv} when we apply ROW–FOLD to S_h.

After we unpack the horizontal intervals, we proceed to unpack the vertical intervals of S_h in a similar fashion. We expand each column of S_h into consecutive columns with each column containing exactly one interval. The way to do the unpacking is slightly different from that for the horizontal intervals. This is because, in general, after unpacking the horizontal intervals some overlapping pair of vertical intervals in adjacent columns may become disjoint. Note that adjacent columns in S_h are not disjoint. However, after unpacking the rows, some adjacent columns may become disjoint (*i.e.* there are no overlapping vertical intervals in the two columns). Our unpacking procedure is as follows: If column i and $i-1$ are not disjoint, we unpack column i as in the case of unpacking the horizontal intervals; if column i and $i-1$ are disjoint, we arbitrarily assign the intervals in column i to the set of consecutive columns. The array obtained after unpacking the columns of S_h is of the form $A_{\pi\tau}$ for some pair of net–permutation π and τ. We now carefully define a partition Ω of τ such that we will get back S_h when we apply COLUMN–FOLD to $A_{\pi\tau}$. Let $\tau = \tau_1\tau_2 \cdots \tau_n$. Recall that column j of $A_{\pi\tau}$ contains the vertical net c_{τ_j}, for all j. Since each column of S_h is expanded into a set of consecutive columns in $A_{\pi\tau}$, the set of nets in each column of S_h is of the form $\{c_{\tau_j}, c_{\tau_{j+1}}, \ldots, c_{\tau_{j+k}}\}$ for some j and k. For every two disjoint adjacent columns in S_h, we partition τ between τ_r and τ_{r+1} where the nets in these two adjacent columns are $\{c_{\tau_a}, c_{\tau_{a+1}}, \ldots, c_{\tau_r}\}$ and $\{c_{\tau_{r+1}}, c_{\tau_{r+2}}, \ldots, c_{\tau_b}\}$. This defines a partition Ω of τ. It is clear that $|\Omega| \leq n$ because it is a partition of $\{1,2,...,n\}$. To see that $|\Omega| \leq m$, we first observe that $|\Omega| = n_1 + 1$ where n_1 is the total number of pairs of disjoint adjacent columns in S_h. Also observe that all disjoint adjacent columns are introduced when we unpack the horizontal intervals of S_{hv}. Since each pair of adjacent horizontal intervals in a row of S_{hv} can at the most contribute to only one pair of disjoint adjacent columns, and there are at most $m-1$ such pairs of adjacent intervals, it follows that $n_1 \leq m-1$. Thus $|\Omega| \leq m$. Again, it is not difficult to see that we will obtain S_h when we apply COLUMN–FOLD to $A_{\pi\tau}$.

Since applying COLUMN–FOLD to $A_{\pi\tau}$ yields S_h, and applying ROW–FOLD to S_h yields S_{hv}, we have $A_{\pi\tau}^{C_\Omega R} = S_{hv}$. Thus, $AREA(A_{\pi\tau}^{C_\Omega R}) = AREA(S_{hv}) \leq AREA(S)$. □

Theorem 8.1 allows us to use the set of all triples (π,τ,Ω) with $|\Omega| \leq \min\{m, n\}$ as the solution space. The cost of a solution (π,τ,Ω) is given by $AREA(A_{\pi\tau}^{C_\Omega R})$.

We now consider an important special case in which all the cells are of uniform size. This case includes structures such as PLA or domino CMOS implementations. In this case, the width of an array is the number of columns and the height is the number of rows. Let β denote the number of columns with only a single cell. The following theorem shows that we can reduce substantially the size of the solution space for most arrays. (For most arrays in practice, β is usually quite small.)

Theorem 8.2. Let A be an unfolded array with uniform cell size. Given any folding solution S, there exists a pair of net–permutations π and τ, and a partition Ω of τ with $|\Omega| \leq \max\{1, \beta\}$, such that $AREA(A_{\pi\tau}^{C_\Omega R}) \leq AREA(S)$.

Proof. Consider the compact folding solution S_{hv} obtained from S. In the case of uniform cell size, the conditions for a compact folding solution are equivalent to the following conditions: (1) every vertical interval in column i overlaps with an interval in column $i-1$, for all $i \geq 2$; and (2) every horizontal interval in row j overlaps with an interval in row $j-1$, for all $j \geq 2$. As before, we first unpack the rows of S_{hv} by expanding each row into a set of consecutive rows such that each row contains exactly one interval.

In this case, any unpacked array S_h obtained by expanding the rows of S_{hv} into consecutive rows has the property that we can recover S_{hv} by applying ROW–FOLD to S_h. We now show how to unpack the rows of S_{hv} to obtain an array S_h so that eventually we can unpack the columns of S_h to obtain an unfolded array $A_{\pi\tau}$ which yields S_h when folded by COLUMN–FOLD with a net-permutation partition Ω of τ such that $|\Omega|$ is small. The idea is to unpack the rows such that overlappings between vertical intervals in adjacent columns are preserved as much as possible. (Recall that the set of horizontal intervals in row i are $J_1^{(i)}, J_2^{(i)}, \ldots, J_{k_i}^{(i)}$ and unpacking the row corresponds to assigning the intervals to a set of consecutive rows p_i+1, p_i+2, ..., p_i+k_i in S_h where $p_i = \sum_{j=1}^{i-1} k_j$.)

Let C be a cell in S_{hv} and let I be the vertical interval containing C. We say C is of type *UP* if C is the bottommost cell of I, is of type *DOWN* if C is the uppermost cell of I, is of type *THRU* if I passes through C, and is of type *ISOLATED* if I is a *degenerate* interval (*i.e.* I consists of the single cell C only).

For each row i, we construct a directed acyclic graph G as follows: corresponding to each interval $J_k^{(i)}$, there is a vertex v_k. (Recall that the intervals $J_k^{(i)}$'s are sorted according to their left end points.) For each pair of vertices v_k and v_{k+1}, let C_1 be the rightmost cell in $J_k^{(i)}$ and C_2 be the leftmost cell in $J_{k+1}^{(i)}$. There is an edge from v_k to v_{k+1} if (1) C_1 and C_2 are in adjacent columns; and (2) (i) C_1 is of type *DOWN*, and C_2 is of type *UP* or *ISOLATED*; or (ii) C_1 is of type *ISOLATED*, and C_2 is of type *UP*. There is an edge from v_{k+1} to v_k if (1) C_1 and C_2 are in adjacent columns; and (2) (i) C_1 is of type *UP*, and C_2 is of type *ISOLATED* or *DOWN*; or (ii) C_1 is of type *ISOLATED*, and C_2 is of type *DOWN*. Let I and J be the two vertical intervals that pass through C_1 and C_2, respectively. Intuitively, if I and J are in adjacent columns and not both intervals are degenerate, an edge from v_k to v_{k+1} (v_{k+1} to v_k) means that we have to place the horizontal interval represented by v_k (v_{k+1}) above the horizontal interval represented by v_{k+1} (v_k) in S_h in order to preserve the overlapping property of the two vertical intervals I and J. A topological labeling of G (with labels from 1 to k_i) corresponds to a particular way to unpack row i of S_{hv}. In fact, if $J_k^{(i)}$ has label j, we assign $J_k^{(i)}$ to row p_i+j of S_h. Suppose S_h is obtained by unpacking the rows of S_{hv} where each row is unpacked by a topological labeling of G. Such an unpacking procedure preserves all of the overlaps between the vertical intervals in adjacent columns except those between degenerate intervals. More precisely, let I and J be two vertical intervals in two adjacent columns of S_{hv}, and let I' and J' be the corresponding intervals in S_h. If not both of I and J are degenerate intervals, we have the following: I and J overlap if and only if I' and J' overlap.

Now we proceed to unpack the vertical intervals of S_h. As in the case where the cell sizes are not uniform, we expand each column of S_h into a set of consecutive columns arbitrarily to obtain an unfolded array $A_{\pi\tau}$ for some pair of net-permutation π and τ. We now define a partition Ω of τ such that the result of COLUMN-FOLD on $A_{\pi\tau}$ is S_h. Let $\tau = \tau_1\tau_2 \cdots \tau_n$. For every pair of adjacent columns in S_h with degenerate vertical intervals in both columns, we partition τ between τ_r and τ_{r+1} where the nets in the two adjacents columns are $\{c_{\tau_a}, c_{\tau_{a+1}}, ..., c_{\tau_r}\}$ and $\{c_{\tau_{r+1}}, c_{\tau_{r+2}}, ..., c_{\tau_b}\}$. This defines a partition Ω of τ. Clearly, there are no adjacent columns with degenerate intervals when β is 0 or 1. In that case, $|\Omega| = 1$. In the case of $\beta > 1$, there are at most β–1 pairs of such adjacent columns, we have $|\Omega| \leq \beta$.

Finally, it is not difficult to see that the result of COLUMN-FOLD on $A_{\pi\tau}$ is S_h, and the result of ROW-FOLD on S_h is S_{hv}. Consequently, $A_{\pi\tau}^{C_\Omega R} = S_{hv}$. Thus, we have $AREA(A_{\pi\tau}^{C_\Omega R}) = AREA(S_{hv}) \leq AREA(S)$. □

8.4.4. The Main Folding Algorithm

We now describe the algorithm SA-FOLD, the simulated annealing based algorithm in [WoLi87] that solves the array optimization problem. The solution space is the set of all possible triples (π,τ,Ω). The cost of a solution (π,τ,Ω) is given by $AREA(A_{\pi\tau}^{C_\Omega R})$. Given a solution (π,τ,Ω), three types of moves are defined.

M1. Swap two values π_i and π_j of the permutation π.
M2. Swap two values τ_i and τ_j of the permutation τ.

M3. Local modification of Ω.

Two solutions (π,τ,Ω) and (π',τ',Ω') are said to be ***neighbors*** if one can be obtained from the other via one of these three moves. M1 corresponds to swapping two rows of $A_{\pi\tau}$, and M2 corresponds to swapping two columns of $A_{\pi\tau}$. There are many possible ways to implement M3. One way is to use a binary sequence ω of length $n-1$ to represent Ω. Let $\omega = \omega_1 \omega_2 \cdots \omega_{n-1}$ be a binary sequence. We can interpret ω_i as a 0 or a 1 to be inserted between τ_i and τ_{i+1}. A 1 between τ_i and τ_{i+1} means they are in different segments. A 0 between τ_i and τ_j means they are in the same segment. For example, suppose $\tau = \tau_1 \tau_2 \tau_3 \tau_4 \tau_5 \tau_6$, and $\omega = 1\,0\,0\,1\,0$. We obtain the sequence $\tau_1 1 \tau_2 0 \tau_3 0 \tau_4 1 \tau_5 0 \tau_6$ after we insert the values of ω into τ. This corresponds to the partition $\Omega = \{ \{\tau_1\}, \{\tau_2, \tau_3, \tau_4\}, \{\tau_5, \tau_6\} \}$. Local modification of Ω can be implemented by local modification of its corresponding binary sequence ω. M3 corresponds to randomly selecting two values ω_i and ω_j of ω and making a probabilistic modification. The size of Ω can be controlled by imposing restrictions on the total number of 1's in ω.

The temperature schedule is of the form $T_k = r \cdot T_{k-1}$, $k = 1,2,3, \cdots$. The value of r is usually between 0.8 and 0.9. The total number of moves (accepted or rejected) allowed for each temperature is equal to N where N is a user specified parameter. The stopping criteria are: (1) the number of accepted moves in that temperature is very small; or (2) the average cost remains unchanged for three consecutive temperatures. The initial solution can be arbitrary. The default initial solution (π,τ,Ω) is given by: $\pi = 1 2 3 \cdots m$, $\tau = 1 2 3 \cdots n$, and $\Omega = \{\{1,2,...,n\}\}$, *i.e.* the given ordering of the nets in A and the trivial partition of the vertical nets are used.

8.5. Experimental Results

The simulated annealing array optimization algorithm was implemented in Pascal on a PYRAMID computer. Experimental results on several test problems are summarized in Tables 8.1 and 8.2. Table 8.1 summarizes the results for the case when all the cells are of uniform size. Table 8.2 summarizes the results for the general case when the array has variable cell sizes. The sparsities of

Example	Uniform Cell Sizes			
	Before folding		After folding	
	mxn	Area	$m_f x n_f$	Area
ex.1	15 x 20	300	9 x 8	72
ex.2	20 x 30	600	17 x 20	340
ex.3	30 x 50	1500	26 x 34	884
ex.4	30 x 40	1200	24 x 27	648
ex.5	25 x 40	1000	14 x 14	196
ex.6	25 x 50	1250	16 x 15	240
ex.7	35 x 60	2100	29 x 35	1015
ex.8	20 x 33	660	11 x 13	143
ex.9	17 x 35	595	10 x 15	150
ex.10	56 x 81	4536	24 x 30	720

Table 8.1

Example (mxn)	Variable Cell Sizes	
	Initial Area	Final Area
ex.11 (15x20)	2763.3	884.4
ex.12 (20x30)	7099.8	4103.5
ex.13 (30x50)	18421.2	11145.9
ex.14 (30x50)	16618.7	5559.7
ex.15 (25x40)	10308.7	3648.7
ex.16 (26x41)	12996.2	6822.5
ex.17 (30x40)	14477.5	7650.2
ex.18 (15x30)	5098.6	2198.8
ex.19 (25x40)	9179.3	2022.7
ex.20 (25x50)	12446.3	2525.2

Table 8.2

the arrays in these examples are between 80% and 90%. The running times for these examples range from 0.5 CPU minutes for ex.1 to 12 CPU minutes for ex.10. Fig.8.7 shows the initial unfolded array and the final folded array of ex.10. The algorithm reduced the number of rows from 56 to 24, and reduced the number of columns from 81 to 30. Finally, Fig.8.8 shows the initial unfolded array and the final folded array of ex.19. The algorithm substantially reduced the area from 9179.3 to 2022.7.

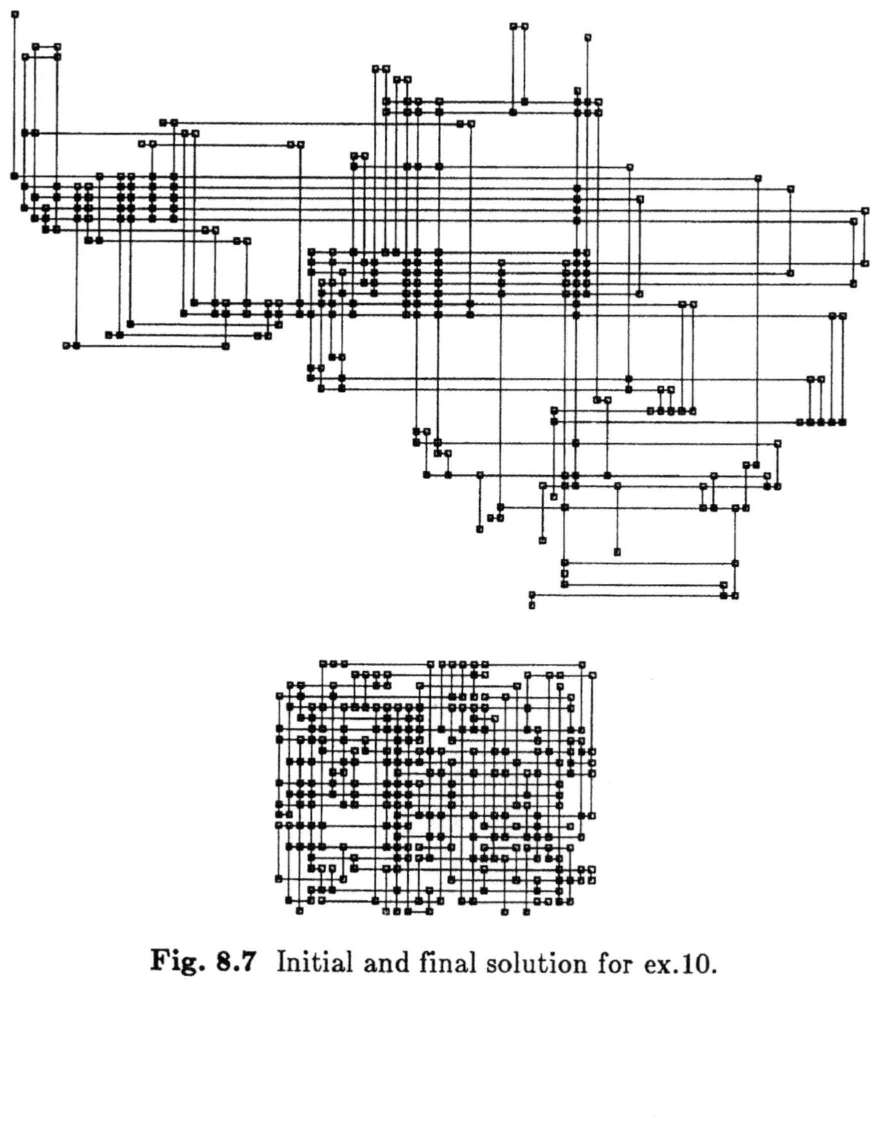

Fig. 8.7 Initial and final solution for ex.10.

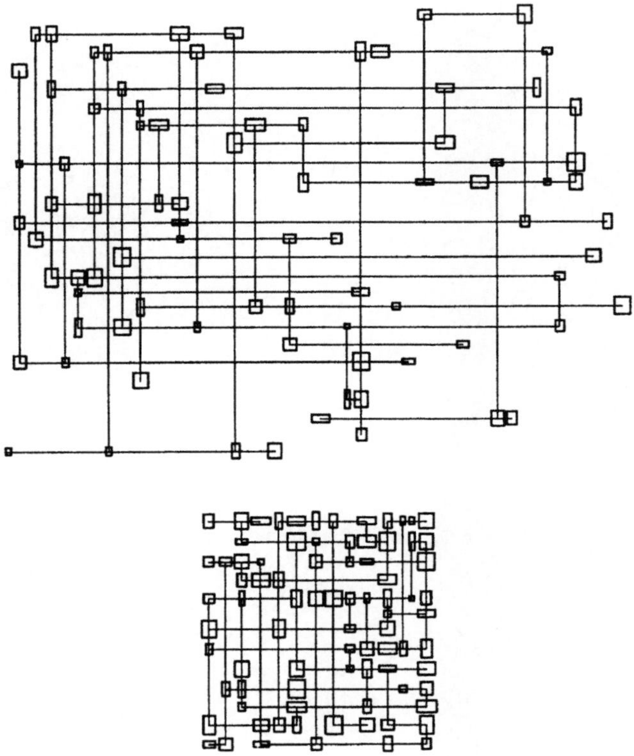

Fig. 8.8 Initial and final solution for ex.19.

8.6. Concluding Remarks

We presented in this chapter a general array optimization algorithm [WoLi87b] based on the method of simulated annealing. The algorithm can be used for compacting Gate Matrix layouts, SLA's, Weinberger Arrays, and for multiple folding of PLA's. Instead of using a solution space consisting of both valid and invalid solutions, the algorithm only use a subset of the valid solutions as the solution space. Futhermore, all optimal solutions are represented in the solution space and it is computationally inexpensive to move from one solution to a neighboring solution. Experimental results for various test problems indicate that the algorithm performs well.

REFERENCES

[AaLa85] Aarts, E. H. L. and P. J. M. van Laarhoven, "Statistical Cooling: A General Approach To Combinatorial Optimization Problems," *Philips Journal of Research,* Vol. 40, (1985), 193–226.

[AhHU74] Aho, A. V., J. E. Hopcroft and J. D. Ullman, *The Design and Analysis of Computer Algorithms,* Addison Wesley, 1974.

[BaJo86] Banerjee, P. and M. Jones, "A Parallel Simulated Annealing Algorithm for Standard Cell Placement on a Hypercube Computer," *Proc. IEEE Intl. Conference on Computer–Aided Design,* (1986), 34–37.

[BaBL83] Baker, B. S., S. N. Bhatt and F. T. Leighton, "An Approximation Algorithm for Manhattan Routing," *Proc. ACM Symp. on Theory of Computing,* (1983), 477–486.

[Breu77] Breuer, M. A., "Min–Cut Placement," *J. Design Automation and Fault–Tolerant Computing,* Vol. 1, No. 4, (1977), 343–362.

[BuPe83] Burstein, M. and R. Pelavin, "Hierarchical Channel Router," *Proc. 20th ACM/IEEE Design Automation Conference,* IEEE (1983), 591–597.

[CaDV85] Catthoor, F., H. DeMan, and J. Vanderwalle, "SAIL-PLANE: A Simulated Annealing based CAD–tool for the Analysis of Limit–Cycle Behavior," *Proc. IEEE Intl. Conference on Computer Design,* (1985), 244–247.

[CaDV86] Catthoor, F., H. DeMan, and J. Vanderwalle, "Investigation of Finite Word–Length Effects in Arbitrary Digital Filters Using Simulated Annealing," *Proc. IEEE Intl. Symp. on Circuits and Systems,* (1986), 1296–1297.

[CaRS86] Casotto, A., F. Romeo, and A. Sangiovanni–Vincentelli "A Parallel Simulated Annealing Algorithm for the Placement of Macro–Cells," *Proc. IEEE Intl. Conference on Computer–Aided Design,* (1986), 30–33.

[Cern85] Cerny, V., "A Thermodynamic Approach to the Traveling Salesman Problem: An Efficient Simulation Algorithm," *J. Opt. Theory and Appl.,* 45, (1985), 41–51.

[Chan83] Chan, W. S., "A New Channel Routing Algorithm," *Proc. 3rd Cal Tech Conference on VLSI,* (1983), 117–139.

[ChRa86] Chung, M. J. and K. K. Rao, "Parallel Simulated Annealing for Partitioning and Routing," *Proc. IEEE Intl. Conference on Computer Design,* (1986), 238–242.

[CoPa86] Cohoon, J. P. and W. D. Paris, "Genetic Placement," *Proc. IEEE Intl. Conference on Computer–Aided Design,* (1986), 422–425.

[DaKN87] Darema, F., S. Kirpatrick, and V. A. Norton, "Parallel Algorithms for Chip Placement by Simulated Annealing," *IBM Journal of Res. Develop.,* Vol. 31, No. 3, (1987), 391–402.

[DeKL87] Deo, N., M. S. Krishnamoorthy, and M. A. Langston, "Exact and Approximate Solutions for the Gate Matrix Layout Problem," *IEEE Trans. on Computer–Aided Design of Integrated Circuits and Systems,* Vol. CAD–6, (1987), 79–84.

[DeNe86a] Devadas, S. and A. Newton, "GENIE: A Generalized Array Optimizer for VLSI Synthesis," *Proc. 23rd ACM/IEEE Design Automation Conference,* (1986), 631–637.

[DeNe86b] Devadas, S. and A. R. Newton, "Topological Optimization of Multiple Level Array Logic: On Uni and Multi–Processors," *Proc. IEEE Intl. Conference on Computer–Aided Design,* (1986), 38–41.

[DeSa83] De Micheli, G. and A. Sangiovanni-Vincentelli, "Pleasure: A Computer Program for Simple/Multiple Constrained/Unconstrainted Folding of Programmable Logic Arrays," *Proc. 20th ACM/IEEE Design Automation Conference,* (1983), 530–537.

[Deut76] Deutsch, D. N., "A 'Dogleg' Channel Router," *Proc. 13th ACM/IEEE Design Automation Conference,* (1976), 425–433.

[DiPi86] Distante, F. and V. Piuri, "Optimal Behavioral Test Procedure for VLSI Devices: A Simulated Annealing Approach," *Proc. IEEE Intl. Conference on Computer Design,* (1986), 31–35.

[Dunh63] Dunham, B. *et al.*, "Design by Natural Selection," *Synthese,* D. Reidel Publ. Comp., Dordrecht Holland, (1963), 254–259.

[DuKe85] Dunlop, A. E. and B. W. Kernighan, "A Procedure for Placement of Standard-Cell VLSI Circuits," *IEEE Trans. on Computer-Aided Design of Integrated Circuits and Systems*, Vol. CAD-4, No. 1, (1985), 92–98.

[Fell50] Feller, W., *An Introduction to Probability Theory and Applications, Vol. 1*, John Wiley & Sons, New York, 1950.

[FlGM85] Fleisher, H., J. Giraldi, D. B. Martin, R. L. Phoenix, and M. A. Tavel, "Simulated Annealing as a Tool for Logic Optimization in a CAD Environment," *Proc. IEEE Intl. Conference on Computer-Aided Design,* (1985), 203–205.

[FiWa87] Fiebrich, R. and C. Wang, "Circuit Placement Based on Simulated Annealing on a Massively Parallel Computer," *Proc. IEEE Conference on Computer Design,* (1987).

[FrKa86] Frankle, J. and R. M. Karp, "Circuit Placements and Cost Bounds by Eigenvector Decomposition," *Proc. IEEE Intl. Conference on Computer-Aided Design,* (1986), 414–417.

[GaJo79] Garey, M. R. and D. S. Johnson, *Computers and Intractability: A Guide to the Theory of NP–Completeness,* W. H. Freeman and Company, San Francisco, 1979.

[GeMi85] Gelfand, S. B. and S. K. Mitter, "Analysis of Simulated Annealing for Optimization," *Proc. 24th Conference on Decision and Control,* (1985), 779–786.

[Golu80] Golumbic, M. C., *Algorithmic Graph Theory and Perfect Graphs,* Academic Press, New York, 1980.

[Gons86] Gonsalves, G., "Logic Synthesis Using Simulated Annealing," *Proc. IEEE Intl. Conference on Computer Design,* (1986), 561–564.

[GoCW84] Gopal, I. S., D. Coppersmith, and C. K. Wong, "Optimal Wiring for Movable Terminals," *IEEE Trans. on Computers,* Vol. C–32, (1983), 845–858.

[GoKu78] Goto, S. and E. Kuh, "An Approach to the Two–Dimensional Placement Problem in Circuit Layout," *IEEE Trans. Circuits and System,* CAS–25, (1978), 208–214.

[Gras82] Grass, W., "A Depth–First Branch and Bound Algorithm for Optimal PLA Folding," *Proc. 19th ACM/IEEE Design Automation Conference,"* (1982), 133–140.

[Gree76] Greer, D. L., "An Associative Logic Matrix," *IEEE Journal of Solid State Circuits,* Vol. SC–11, (1976), 679–691.

[Grov86] Grover, L. K., "A New Simulated Annealing Algorithm for Standard Cell Placement," *Proc. IEEE Intl. Conference on Computer–Aided Design,* (1986), 378–380.

[Grov87] Grover, L. K., "Standard Cell Placement using Simulated Sintering," *Proc. 24th ACM/IEEE Design Automation Conference,* (1987), 56–59.

[GrSu84] Greene, J. W. and K. J. Supowit, "Simulated Annealing without Rejected Moves," *Proc. IEEE Intl. Conference on Computer Design*, (1984), 658–663.

[Haje85] Hajek, B., "Cooling Schedule for Optimal Annealing," *Mathematics of Operations Research*, (1987), to appear.

[HaNS82a] Hachtel, G. D., A. Newton, and A. Sangiovanni–Vincentelli, "An Algorithm for Optimal PLA Folding," *IEEE Trans. on Computer–Aided Design of Integrated Circuits and Systems,"* Vol. CAD–1, (1982), 62–76.

[HaNS82b] Hachtel, G. D., A. Newton, and A. Sangiovanni–Vincentelli, "Techniques for Programmable Logic Array Folding," *Proc. 19th ACM/IEEE Design Automation Conference,* (1982), 147–155.

[HaSt71] Hashimoto, A., and J. Stevens, "Wire Routing by Optimizing Channel Assignment within Large Apertures," *Proc. 8th ACM/IEEE Design Automation Workshop,* (1971), 214–224.

[HaWA73] Hansan M., P. K. Wolf, and B. J. Anguli, "Some Experimental Results on Placement Techniques," *Proc. 13th Design Automation Conference,* (1973), 214–224.

[Holl75] Holland, J. H., *Adaptation in Natural Artificial Systems,* University of Michigan Press, 1975.

[HsLL87] Hsieh, T. M., H. W. Leong, and C. L. Liu, "Two–Dimensional Compaction by Simulated Annealing," to appear.

[HuRS86] Huang, M. D., F. Romeo, and A. Sangiovanni–Vincentelli, "An Efficient General Schedule for Simulated Annealing," *Proc. IEEE Intl. Conference on Computer–Aided Design,* (1986), 381–384.

[HwFK86] Hwang, D. K., W. K. Fuchs, and S. M. Kang, "An Efficient Approach to Gate Matrix Layout," *Proc. IEEE Intl. Conf. on Computer–Aided Design*, (1986), 312–315.

[JeGe83] Jepsen, D. W. and C. D. Gelatt, "Macro Placement by Monte Carlo Annealing," *Proc. IEEE Intl. Conference on Computer Design,* (1983), 495–498.

[KaFu79] Kashiwabara, T. and T. Fujisawa, "An NP-complete Problem on Interval Graphs," *Proc. IEEE Symp. on Circuits and Systems,* (1979), 82–83.

[KeLi70] Kernighan, B. W. and S. Lin, "An Efficient Heuristic Procedure for Partitioning Graphs," *Bell Systems Technical Journal,* 49, No. 2, (1970), 291–307.

[KeSP73] Kernighan, B. W., D. G. Schweikert, and G. Persky, "An Optimal Channel-Routing Algorithm for Polycell Layouts of Integrated Circuits," *Proc. 10th ACM/IEEE Design Automation Workshop,* (1973), 50–59.

[KiGV83] Kirkpatrick, S., C. D. Gelatt, and M. P. Vecchi, "Optimization by Simulated Annealing," *Science,* Vol. 220, (1983), 671–680.

[KlBa87] Kling, R. M. and P. Banerjee, "ESP: A New Standard Cell Placement Package using Simulated Evolution," *Proc. 24th ACM/IEEE Design Automation Conference,* (1987), 60–66.

[KoDr84] Kobayashi, H. and C. E. Drozd, "Efficient Algorithms for Routing Interchangeable Terminals," *Proc. IEEE Intl. Conference on Computer Design,* (1984), 79–81.

[KoFo84] Kobayashi, H. and Y. P. Foo, "Programmable Logic for Parallel Convolution," *Proc. IEEE Intl. Conference on Computer Design,* (1984), 700–704.

[KrRu87] Kravitz, S. A. and R. A. Rutenbar, "Placement by Simulated Annealing on a Multiprocessor," *IEEE Trans. on Computer-Aided Design of Integrated Circuits and Systems*, Vol. CAD-6, No. 4, (1987), 534–549.

[Laut80] Lauther, U., "A Min-Cut Placement Algorithm for General Cell Assemblies Based on a Graph Representation," *Journal of Digital Systems,* Vol. IV, Issue 1, (1980), 21–34.

[Lawl76] Lawler, E. L., *Combinatorial Optimization*, Holt, Rinehart & Winston, New York, 1976.

[LaAa87] Larhoven, P. J. M. van and E. H. L. Aarts, *Simulated Annealing: Theory and Applications*, Kluwer Academic Publishers, Boston, 1987.

[LaDe86] Lam, J. and J. M. Delosme, "Logic Minimization Using Simulated Annealing," *Proc. IEEE Intl. Conference on Computer-Aided Design*, (1986), 348–351.

[LaDi85] LaPotin, D. P. and S. W. Director, "Mason: A Global Floor-Planning Tool," *Proc. IEEE Intl. Conference on Computer-Aided Design,* (1985), 143–145.

[Leon86a] Leong, H. W., "Routing Problems in Physical Design of Integrated Circuits," *Ph.D. Thesis, Department of Computer Science, University of Illinois at Urbana-Champaign,* 1986.

[Leon86b] Leong, H. W., "A New Algorithm for Gate Matrix Layout," *Proc. IEEE Intl. Conference on Computer-Aided Design*, (1986), 316–319.

[LeLi83] Leong, H. W. and C. L. Liu, "A New Channel Routing Problem," *Proc. 20th ACM/IEEE Design Automation Conference,* (1983), 584–590.

[LeLi85] Leong, H. W., and C. L. Liu, "Permutation Channel Routing," *Proc. IEEE Intl. Conference on Computer Design*, (1985), 579–584.

[LeLi87] Leong, H. W. and C. L. Liu, "Algorithms for Permutation Channel Routing," *INTEGRATION, the VLSI journal, 5 (1987), 17–45.*

[LeWL85] Leong, H. W., D. F. Wong, and C. L. Liu, "A Simulated-Annealing Channel Router," *Proc. IEEE Intl. Conference on Computer-Aided Design,* (1985), 226–228.

[LeLi84] Lewandowski, J. L. and C. L. Liu, "A Branch and Bound Algorithm for Optimal PLA Folding," *Proc. 21st ACM/IEEE Design Automation Conference,*

(1984), 426–433.

[Li83] Li, J. T. "Algorithms for Gate Matrix Layout," *Proc. IEEE Intl. Symp. on Circuits and Systems*, (1983), 1013–1016.

[LiAB86] Ligthart, M. M., E. H. L. Aarts, and F. P. M. Beenker, "Design–For–Testability of PLA's Using Statistical Cooling," *Proc. 23rd ACM/IEEE Design Automation Conference,* (1986), 339–345.

[LiKe73] Lin, S. and B. W. Kernighan, "An Effective Heuristic for the Travelling–Salesman Problem," *Operations Research,* 21, (1973), 498–516.

[LoLa80] Lopez, A. D. and H. F. S. Law, "A Dense Gate Matrix Layout Method for MOS VLSI," *IEEE Trans. on Electronic Devices,* Vol. ED–27, No. 8, (1980), 1671–1675.

[LuMe86] Lundy, M. and A. Mess, "Convergence of an Annealing Algorithm," *Math. Prog.,* 34, (1986), 111–124.

[LuVV82] Luby, M., U. Vazirani, V. Vazirani and A. Sangiovanni–Vincentelli, "Some Theoretical Results on the Optimal PLA Folding Problem," *Proc. IEEE Intl. Circ. and Comp. Conference,* (1982), 165–170.

[MaMH82] Maling, K., S. H. Mueller, and W. R. Heller, "On Finding Most Optimal Rectangular Package Plans," *Proc. 19th ACM/IEEE Design Automation Conf.,* (1982), 663–670.

[MeCo80] Mead, C. and L. Conway, *Introduction to VLSI Systems.* Addison–Wesley, New York, 1980.

[MeRR53] Metropolis, N., A. Rosenbluth, M. Rosenbluth, A. Teller, and E. Teller, "Equation of State Calculations by Fast Computing Machines," *Journal of Chemical Physics,* 21, (1953), 1087–1092.

[MiRS86] Mitra, D., F. Romeo, and A. Sangiovanni–Vincentelli, "Convergence and Finite–Time Behavior of Simulated Annealing," *Advances of Applied Probaility,* 18, (1986), 747–771.

[MiVa80] Micali, S. and V. Vazirani, "An $O(\sqrt{V} \cdot E)$ Algorithm for Finding Maximum Matching in General Graphs," *Proc. 21st Annual IEEE Symposium on Foundations of Computer Science,* (1980), 17–27.

[MoGe85] Moore, T. P. and A. J. de Geus, "Simulated Annealing Controlled by a Rule–Based Expert System," *Proc. IEEE Intl. Conference on Computer–Aided Design,* (1985), 200–202.

[Moli85] Molitor, P., "Layer Assignment by Simulated Annealing," *Technical Report, University of Saarlandes, West Germany,* 1985.

[Most86] Mosteller, R. C., *Ph.D. Thesis, California Institute of Technology, Pasadena,* 1986.

[OhMK79] Ohtsuki, T., H. Mori, E. Kuh, O. Kashiwabara, and T. Fujisawa, "One–Dimensional Logic Gate Assignment and Interval Graphs," *IEEE Trans. on Circuits and Systems,* Vol. CAS–26, No. 9, (1979), 675–684.

[Osma87] Osman, W., "Two–dimensional Compaction of Abstract Layouts with Statistical Cooling," *Philips Research Report,* 1987.

[Otte82] Otten, R. H. J. M., "Automatic Floorplan Design," *Proc. 19th ACM/IEEE Design Automation Conference,* (1982), 261–267.

[Otte83] Otten, R. H. J. M., "Efficient Floorplan Optimization," *Proc. IEEE Intl. Conference on Computer Design,* (1983), 499–502.

[OtGi84] Otten, R. H. J. M., L. P. P. P. van Ginneken, "Floorplan Design using Simulated Annealing," *Proc. IEEE Intl. Conference on Computer–Aided Design,* (1984), 96–98.

[PaSt82] Papadimitriou, C. H. and K. Steiglitz, *Combinatorial Optimization: Algorithms and Complexity,* Prentice Hall, New York, 1982.

[PiDe86] Pincus, J. P. and A. M. Despain, "Delay Reduction Using Simulated Annealing," *Proc. 23rd ACM/IEEE Design Automation Conference,* (1986), 690–695.

[PrCh85] Preas, B. and C. S. Chow, "Placement and Routing Algorithms for Topological Integrated Circuit Layout," *Proc. IEEE Intl. Symp. on Circuits and Systems,* (1985), 17–20.

[PrVa79] Preas, B. and W. M. VanCleemput, "Placement Algorithms for Arbitrary Shaped Blocks," *Proc. 16th ACM/IEEE Design Automation Conference,* (1979), 474–480.

[QuBr79] Quin, N. R. and M. A. Breuer, "A Force Directed Component Placement Procedure for Printed Circuit Boards," *IEEE Trans. on Circuit and Systems*, Vol. CAS–26, No. 6, (1979), 377–388.

[RaPa87] Ravikumar, C. P. and L. M. Patnaik, "Parallel Placement Based on Simulated Annealing," *Proc. IEEE Intl. Conference on Computer Design,* (1987).

[ReND77] Reingold, E. M., J. Nievergelt, and N. Deo, *Combinatorial Algorithms: Theory and Practice*, Prentice Hall, New York, 1977.

[ReSS85] Reed, J., A. Sangiovanni–Vincentelli and M. Santomauro, "A New Symbolic Channel Router: YACR2," *IEEE Trans. on Computer–Aided Design of Integrated Circuits and Systems,* Vol. CAD–4, (1985), 208–219.

[RiFi82] Rivest, R. L. and C. M. Fiduccia, "A 'Greedy' Channel Router," *Proc. 19th ACM/IEEE Design Automation Conference,* (1982), 418–424.

[RoBS86] Rose, J. S., D. R. Blythe, W. M. Snelgrove, and Z. G. Vranesic, "Fast, High Quality VLSI Placement on an MIMD Multipsrocessor," *Proc. IEEE Intl. Conference on Computer–Aided Design,* (1986), 42–45.

[RoSa85] Romeo, F. and A. Sangiovanni–Vincentelli, "Probabilistic Hill Climbing Algorithms: Properties and Applications," *Proc. 1985 Chapel Hill Conference on*

VLSI, (1985), 393–417.

[SeSa85] Sechen, C. and A. Sangiovanni–Vincentelli, "The Timberwolf Placement and Routing Package," *IEEE Journal of Solid–State Circuits,* Vol. SC–20, No. 2, (1985), 510–522.

[SeSa86] Sechen, C. and A. Sangiovanni–Vincentelli, "TimberWolf 3.2: A New Standard Cell Placement and Global Routing Package," *Proc. 23rd ACM/IEEE Design Automation Conference,* (1986), 432–439.

[ShDu85] Sha, L. and R. W. Dutton, "An Analytical Algorithm for Placement of Arbitrary Sized Rectangular Blocks," *Proc. 22nd ACM/IEEE Design Automation Conference,* (1985), 602–608.

[Skri85] Skrien, D. J., "An Algorithm for Optimal Multiple Foldings of Programmable Logic Arrays," *M.S. Thesis, Department of Computer Science, University of Illinois at Urbana–Champaign,* 1985.

[Souk81] Soukup, J., "Circuit Layout," *Proc. IEEE,* Vol. 69, (1981), 1281–1304.

[Stev72] Stevens, J., "Fast Heuristic Techniques for Placing and Wiring Printed Circuit Boards," *Ph.D. Thesis, University of Illinois at Urbana–Champaign,* 1972.

[Stoc83] Stockmeyer, L., "Optimal Orientations of Cells in Slicing Floorplan Designs," *Information and Control,* Vol. 59, (1983), 91–101.

[SuKe86] Suaris, P. R. and G. Kedem, "Standard Cell Placement by Quadrasection," *MCNC Technical Report TR86–14,* 1986.

[Szym85] Szymanski, T., "Dogleg Channel Routing is NP–Complete," *IEEE Trans. on Computer–Aided Design of Integrated Circuits and Systems,* Vol. CAD–4, (1985), 31–41.

[Tarj83] Tarjan, R. E., *Data Structures and Network Algorithms,* CBMS–NSF Regional Conference Series in

Applied Mathematics, 44, SIAM, 1983.

[VeKi83] Vecchi, M. P. and S. Kirkpatrick, "Global Wiring by Simulated Annealing," *IEEE Trans. on Computer-Aided Design of Integrated Circuits and Systems,* Vol. CAD–2, (1983), 215–222.

[Wein67] Weinberger, A., "Large–scale Integration of MOS Complex Logic: A Layout Method," *IEEE Journal of Solid-State Circuits,* Vol. SC–2, (1967), 182–190.

[Whit84] White, S. R., "Concepts of Scale in Simulated Annealing," *Proc. IEEE Intl. Conference on Computer Design,* (1984), 646–651.

[Wing82] Wing, O. "Automated Gate Matrix Layout," *Proc. IEEE Intl. Symp. on Circuits and Systems,* (1982), 681–685.

[Wing83] Wing, O. "Interval Graph Based Gate Matrix Layouts," *Proc IEEE Intl. Conf. on Computer-Aided Design,* (1983), 84–85.

[WiHW85] Wing, O., S. Huang, and R. Wang, "Gate Matrix Layout," *IEEE Trans. on Computer-Aided Design of Integrated Circuits and Systems,* Vol. CAD–4, No. 3, (1985), 220–231.

[Wong87] Wong, D. F., "Algorithmic Aspects of VLSI Circuit Layout," *Ph.D. Thesis, Department of Computer Science, University of Illinois at Urbana–Champaign,* 1987.

[WoLi86] Wong, D. F. and C. L. Liu, "A New Algorithm for Floorplan Design," *Proc. 23rd ACM/IEEE Design Automation Conference,* (1986), 101–107.

[WoLi87a] Wong, D. F. and C. L. Liu, "Floorplan Design for Rectangular and L–shaped Modules," *Proc. IEEE Intl. Conference on Computer-Aided Design,* (1987), 520–523.

[WoLi87b] Wong, D. F. and C. L. Liu, "Array Optimization for VLSI Synthesis," *Proc. 24th ACM/IEEE Design Automation Conference,* (1987), 537–543.

[WoLL86] Wong, D. F., H. W. Leong and C. L. Liu, "Multiple PLA Folding by the Method of Simulated Annealing," *Proc. IEEE Custom Integrated Circuits Conference,* (1986), 351–355.

[WoLL87] Wong, D. F., H. W. Leong, and C. L. Liu, "PLA Folding by Simulated Annealing," *IEEE Journal of Solid State Circuits,* Vol. SC–22, No 2, (1987), 208–215.

[WoWT86] Woo, L. S., C. K. Wong, and D. T. Tang, "Pioneer: A Macro–Based Floor–Planning Design System," *VLSI SYSTEMS DESIGN,* (August, 1986), 32–43.

[YaLi87] Yao, X. J. and C. L. Liu, "PLA Logic Minimization by Simulated Annealing," to appear.

[YoKu82] Yoshimura, T. and E. S. Kuh, "Efficient Algorithms for Channel Routing," *IEEE Trans. on Computer–Aided Design of Integrated Circuits and Systems,* Vol. CAD–1, (1982), 25–35.

[YuWi85] Yu, Q. and O. Wing, "Interval–graph–based PLA Folding," *INTEGRATION, the VLSI journal,* Vol. 3, (1985), 33–48.

[Zees85] Zeestraten, R. J. A., "Two–Dimensional Compaction," *Philips Research Report,* 1985.

INDEX

GPSR Compliance

The European Union's (EU) General Product Safety Regulation (GPSR) is a set of rules that requires consumer products to be safe and our obligations to ensure this.

If you have any concerns about our products, you can contact us on ProductSafety@springernature.com

In case Publisher is established outside the EU, the EU authorized representative is:

Springer Nature Customer Service Center GmbH
Europaplatz 3
69115 Heidelberg, Germany

Batch number: 10682323

Printed by Printforce, the Netherlands